AMERICAN PHILOSOPHY OF TECHNOLOGY

The Indiana Series in the Philosophy of Technology

Don Ihde, general editor

AMERICAN PHILOSOPHY OF TECHNOLOGY: THE EMPIRICAL TURN

Edited by
Hans Achterhuis

Translated by
Robert P. Crease

INDIANA UNIVERSITY PRESS
Bloomington and Indianapolis

The translation of this book was made possible by generous financial support from the Department of Philosophy, Twente University

This book is a publication of
Indiana University Press
601 North Morton Street
Bloomington, IN 47404-3797 USA
http://iupress.indiana.edu

Telephone orders   800-842-6796

Fax orders   812-855-7931

Orders by e-mail   iuporder@indiana.edu

Published in Dutch as *Van stoommachine tot cyborg: Denken over techniek in de nieuwe wereld*, edited by Hans Achterhuis. ©1997 by Uitgeverij Ambo bv, Amsterdam

The paper used in this publication meets the minimum requirements of American National Standard for Information Sciences—Permanence of Paper for Printed Library Materials, ANSI Z39.48-1984. Manufactured in the United States of America
CIP and print line to come

**Library of Congress
Cataloging-in-Publication Data**

Van stoommachine tot cyborg. English
    American philosophy of technology : the empirical turn / edited by Hans Achterhuis ; translated by Robert P. Crease.
        p.   cm — (Indiana series in the philosophy of technology)
    Includes bibliographical references and index.
    ISBN 0-253-33903-0 (cl : alk. paper) — ISBN 0-253-21449-1 (pa : alk. paper)
    1. Technology—Philosophy.   I. Achterhuis, Hans, date II. Title.   III. Series.

T14 .V34 2001
601—dc21
                                        00-047138
  1   2   3   4   5   06   05   04   03   02   01

CONTENTS

Foreword

Has the center of gravity for front-rank work in the philosophy of technology shifted from Europe to North America? That seems to be one of the implications to be drawn from this new book on contemporary American philosophy of technology. And what is perhaps of most interest is that the studies presented here were prepared by Europeans, more precisely, by six Dutch philosophers.

The Netherlands has had a long and deep tradition in the philosophy of technology, and those of us who have been familiar with the Dutch and their ways of dealing with technologies—and the environment—have appreciated their democratic and considered ways of social development. Technology assessment in particular is a process that involves both an educated populace and deliberate democratic discussion. In contemporary times the decisions to turn 20 percent of the lands once claimed from the North Sea into "nature" parks, and *not* to go ahead with breeding processes made possible by "Herman," a genetically altered bull designed to lower lactose allergies, were made through public discussion and evaluation. Such results might seem unusual in the North American context, but precisely because the Dutch public is so deliberately involved with such decisions, its voice is mandated therein.

This book presents a European perspective on contemporary American philosophy of technology. Even more unique is the route by which this book arrived at its published state. In 1993, Hans Achterhuis, a prominent Dutch philosopher of technology, edited and published *De maat van de techniek* (literally, *The Matter of Technology*), with the help of several of his colleagues at Twente University. This earlier book could be called a study of the "classical" founders of philosophy of technology, including Lewis Mumford, Martin Heidegger, Jacques Ellul and Hans Jonas, all notable as the initiators of philosophical thinking about technology. *De maat van de techniek* proved to be both an insightful survey into the founders' generation and a useful upper-level introduction to the major themes of philosophy of technology.

In 1997, Achterhuis, with five of his philosophy department colleagues, published *Van stoommachine tot cyborg: Denken over techniek in de nieuwe wereld* (literally, *From Steam Engine to Cyborg: Thinking Technology in the New World*). After struggling through the introduction in Dutch (thus calling upon me to transpose as best I could from a German reading competence), I realized that this book not only brought philosophy of technology up to date from its early-twentieth-century origins, but it did so through the respected perspective of European scholars looking at what they regard as the contemporary state of the art. Our Dutch colleagues see, perhaps better than do we in the "New World," what transformations have occurred in this "emigration" to North America. They have articulated what North American philosophers might well wish to acknowledge, but they do so with the greater authority of a distanced perspective.

The reader should take careful note of the introduction, which lays out the differences, for example, between the high-altitude and "transcendental" perspectives of our acknowledged "godfathers" and the lower-altitude, more particular and pragmatic looks at technologies of the Americans included here. This is what Achterhuis calls an "empirical turn." The tone of this generation is different, too. Whereas the "godfathers" were often dystopian, the Americans' outlook is more ambivalent and reflects greater optimism concerning technologies. And the earlier, totalizing term "Technology" (*überhaupt*) is replaced by "technologies" in their relational and contextual implications. Of course the twentieth century's world wars and the development of new technologies, particularly information technologies, have changed the historical context. Nevertheless, the undercurrent here is that newer critiques and interpretations are called for, and that these are being provided by the philosophers of technology whose work is examined here.

Achterhuis and his colleagues each prepared a chapter on an individual philosopher whose writings represent an original aspect of the "empirical turn" in philosophy of technology and who has had a significant influence on its current direction. Each author sets the philosopher's work within the context of his or her full development, often illustrating shifts from earlier to more recent interests. Thus we see a spectrum of philosophical development, clearly and concisely presented.

One other aspect of the book's preparation needs to be mentioned here. During the time it took for this English-language edition to be organized and translated (by Robert P. Crease, a philosopher of science and technology who learned Dutch while at the Husserl Archives as a graduate student), three of the selected philosophers had published new books not discussed in the original, 1997 Dutch edition. The authors therefore revised and updated their chapters to take account of these most recent directions. Fortunately, contemporary technologies, primarily electronic, facilitated the whole process of communication and exchange of ideas among the translator, the Dutch authors, and the American philosophers while work on the book was in progress.

The result is not only an overview of the work of six American philosophers of technology and the themes and problems with which each deals, but also a clear, concise introduction to contemporary philosophy of technology. Used in conjunction with selections from the authors discussed, this book provides an ideal text for an entire course on this topic. More generally, it offers an insightful and readable introduction to a relatively new subfield of philosophy.

DON IHDE, *Series Editor*

AMERICAN PHILOSOPHY OF TECHNOLOGY

Introduction: American Philosophers of Technology

Hans Achterhuis

> Again, we should notice the force, effect, and consequences of inventions, which are nowhere more conspicuous than in those three which were unknown to the ancients; namely, printing, gunpowder, and the compass. For these three have changed the appearance and state of the whole world: first in literature, then in warfare, and lastly in navigation; and innumerable changes have been thence derived, so that no empire, sect, or star, appears to have exercised a greater power and influence on human affairs than these mechanical discoveries.
>
> —FRANCIS BACON, *Novum Organum*

This classic text, which was repeated and paraphrased innumerable times in the seventeenth and eighteenth centuries to describe the cultural and social impact of newly discovered technologies, can serve as the motto for the six portraits of philosophers of technology included in this book. All six philosophers emphasize the power of contemporary technology to transform reality. All six interpret and analyze philosophically the profound ways in which technology has transformed—and continues to transform—social networks and forms of life, human wants and possibilities, and the experience of our bodies and of nature.

Let me try to convey the flavor of the themes discussed in this book by mentioning my own experience of how contacts between the United States and my own country (The Netherlands) have been radically transformed over the course of the past few decades. At the beginning of the 1960s, when I first traveled to the United States, I did so by boat. The trip took nine days. I participated in the extremely relaxing shipboard social life, with plenty of reading time, games to play, gossiping, hanging out, and flirting at organized dance events—an atmosphere captured again and again in novels from the first half of the twentieth century. During the crossing, a frequent topic of conversation was the amazing "hot line" recently established between Washington and Moscow. Kennedy and Khrushchev could now talk to each other directly by telephone—whenever necessary, day or night!—and we joked about how much it would cost. At that time, a short international phone call might easily cost a hundred dollars, and calling the United States from The Netherlands was thus only for the rich. I felt privileged to the point of extravagance when, shortly after being picked up by my American hosts, I was allowed to make a short call home. This was the only time that I phoned home during my stay, and all further contact with The Netherlands took place only via thick letters that took approximately a week to reach their destination.

Contrast that with the exchanges between The Netherlands and the United States that took place during the preparation of this book. Thanks to the Internet, we had direct contact with most of the six thinkers profiled here. Don Ihde provided details about his relation to Heidegger, Langdon Winner furnished biographical information, Andrew Feenberg made extensive elaborations of his positions, and Albert Borgmann even dropped by to converse with us and with students in person. In effect, these Americans were closer to us than our Dutch colleagues who lacked Internet access; it was often easier and quicker to get a discussion going with the Americans than with philosophers from Maastricht or Amsterdam.

Still more dramatic contrasts between European and American contacts of the past and present could be made, of course, by using earlier benchmarks: the voyage of Columbus, of the Mayflower, of the first crossing of the Atlantic by a steamboat in 1838 (a passage that took eighteen days and ten hours), or of Lindbergh's solo flight across the Atlantic in May 1927. Another part of this story is the changing cost and number of transatlantic flights and telephone conversations. In 1950, for instance, a ten-minute phone call from the United States to England cost $209.30; by 1989 this had dropped to $10.00. Given this steep reduction, it is not surprising that the number of telephone conversations made annually during this period jumped from 110,300 to almost 85 million.

CLASSICAL PHILOSOPHY OF TECHNOLOGY

Enough anecdotes and facts. Clearly, Bacon's observation about the transforming impact of technology, made at the beginning of the seventeenth century, is as topical as ever. Technological developments have profoundly altered the nature of the exchanges between The Netherlands and the United States, as well as the human experiences that accompany them. These developments have turned upside down our ideas about vacationing and speed, and about what's far and what's near, and have significantly modified social, political, and economic relations. The six thinkers profiled in this collection seek to explore the depths and the extent of just these kinds of transformations in our technological culture.

This was already a concern of the first-generation philosophers of technology, thinkers like Martin Heidegger, Hans Jonas, and Jacques Ellul—who were also fond of citing Bacon's famous remark. Still, these "classical" philosophers of technology foresaw only incompletely, at best, the challenges faced by the thinkers profiled in this book. The classical philosophers of technology occupied themselves more with the historical and transcendental conditions that made modern technology possible than with the real changes accompanying the development of a technological culture. Their approach produced valuable insights. Still, these insights were necessarily circumscribed, because the approach to the technological relation to reality when one is looking at its conditions of possibility leaves unopened many areas of inquiry that can be pursued when one begins to look at the manifold ways in which technology manifests itself.

This latter approach, which looks at the concrete empirical manifestations of different technologies, is precisely the one adopted and explored by contemporary American philosophers of technology. By doing so they are by no means breaking with the "founding fathers" of this philosophical discipline. Rather, the new directions in which most of these thinkers set out are possible only because the preceding generation has transformed the landscape in which they move. Or, to change the image, the six American philosophers of technology profiled in this collection are able to achieve surprising new perspectives because they stand on the shoulders of the giants of classical philosophy of technology. This is not to imply, of course, that these thinkers are dwarfs; as the reader will quickly discover, their contributions are far too great and original.

As philosophical pioneers, the classical philosophers of technology understood that what Ellul called *l'enjeu du siècle* (the challenge of our century) must be sought in modern technology. In confronting this challenge, they made discoveries and posed questions that marked out the path that would eventually lead to an empirically oriented philosophy of technology. These discoveries include the idea that technology must not be thought of as applied natural science, that it is less an instrument than a form of life, and that it must be understood as a "system" (in Ellul's word) or as a "megamachine" (Mumford).

The most important discovery of classical philosophy of technology is undoubtedly the absolute novelty, within the history of humanity, of the technological approach to reality. In "the general pattern," the phrase with which the Dutch historian Jan Romein characterizes all pre-modern cultures, an innumerable and endlessly varying set of symbolic-linguistic relations stands at the center of the human condition. For however much a human being in traditional cultures was a *homo faber*—an artisan or mechanic—technological activities were always guided by the symbolic cultural reality that preceded them and in which they were embedded. Modern technology broke down this hierarchy. Bacon and Descartes already called attention to this totally new, no longer symbolically driven approach to reality in a number of well-known passages. In Bacon's *New Atlantis*, the technologists of "Solomon's House" were charged with, among other things, "enlarging of the bounds of Human Empire, to the effecting of all things possible." And Descartes speaks in analogous terms about the possibility of attaining knowledge useful to life and of applying natural phenomena "in all those uses to which they are adapted and thus render ourselves the masters and possessors of nature." This purely technological approach to reality does not recognize any cultural or other constraints that might arise from the human condition.

It is the great merit of the classical philosophy of technology to have perceived this as amounting to a new and radical turning point in the history of humanity. The questions it posed about this development, and its efforts to understand modern technology as "the other" of the symbolic-linguistic approach to reality, continue to guide the philosophy of technology.

But the new generation of philosophers of technology have provided new answers to these same questions. Mitcham, the most important historian of the philosophy of technology, postulates that the crucial question that humanity has to answer in relation to technology concerns the choice between two theories of human nature (Mitcham 1994). One theory sees a human being as essentially a *homo faber*, a productive being; the other as a *zoön logon echon*, a *homo loquax*, a being characterized by the linguistic. The classical philosophers of technology always correctly noted a sharp gap between these two approaches. However, they ultimately shrank back from acknowledging how fundamental this gap was, and sought to place *homo faber* and the technological approach again hierarchically beneath the sway of the linguistic. They thus chose to pursue the second approach and shrank back from the challenge that Bacon and Descartes had been the first to pose.

Heidegger pointedly formulated the challenge posed by technology to philosophy by arguing that "the essence of modern technology is not something human," and that "modern technology" is much more "than a simple manifestation of human culture." However, after his radical but all-too-brief examination of the "not-human" of technology, Heidegger turned away from it toward language. Only out of language, he claimed, can the essence of technology as the highest danger for human being be thought; only out of it, too, can a possible saving power arise. After Heidegger describes the gap between the linguistic and the technological conceptions, he sees a retreat back into properly linguistic terrain as the only possibility for him as a philosopher. From out of it, he claims, technology is understood, ordered, and bounded.

This Heideggerian move is repeated in its main outlines in the work of all of the classical philosophers of technology. Ellul, who continually thematized the radical and irreducible difference between language and technology, goes so far as to deny the possibility of a technological culture and even of the possibility of a philosophical thinking about technology. Most of the others see this relation as a hierarchical subordination of technology to language. In the name of what Jonas calls "the image of man," technology has to be limited and guided by pregiven cultural values and norms. Arendt dislikes the ambition of technological science "to escape the human condition" instead of being symbolically proportional to it. For her, the great danger is that scientists and technologists "move in a world where speech has lost its power." Technological knowledge and thinking have, in her words, "parted company for good" (Arendt 1958, 3). Her warning seemed to be that, if we do not want human beings to become powerless slaves of technology, we have to subject technology to language and to thinking. Illich, like Heidegger, sees the greatest danger in computer language that he considers an Orwellian newspeak. If human beings are deprived of their natural language, they thereby also lose their essence (Illich and Sanders 1988 , 106).

All of these approaches fathom the essence of modern technology more deeply than the customary descriptions that portray technology as an instrument or as one cultural expression among others. At the same time, however, they refrain from addressing concrete technological practices and developments, and fail as well to appreciate how these can rapidly alter the actual normative frameworks of culture. But this is precisely the main issue in the contemporary challenge of technology to philosophical thinking.

In an interview with the journal *Nachrichten aus Chemie, Technik, und Laboratorium* (Jonas 1985, 301–23), Jonas once set forth, as the practical consequence of his philosophical observations about technology, a list of what should not be allowed to be developed technologically. In response to the question, "You have said what science must keep its fingers out of. . . . Can you, as a philosopher, provide scientists with specific indications as to which ways and in which areas to work, accompanied by a positive, fleshed-out ethics?" Jonas answered that he had not yet thought that through. "I admit that I am not prepared to answer that question. . . . You are completely right that, up to the present, my efforts have been directed at investigating where we must say no, or where we must hold back" (313). This answer, in broad outline, illustrates the approach of the classical philosophers of technology. They speak about a priori boundaries, restrictions, prohibitions in the technological approach to "the effecting of all things possible."

THE EMPIRICAL TURN

About two decades ago, dissatisfaction with the existing, classical philosophical approach to technology among those who studied new developments in technological culture as well as the design stages of new technologies led to an empirical turn that might roughly be characterized as constructivist. This empirical turn was broader and more diverse than the one that had taken place earlier in the philosophy of science, especially as inspired by the work of Thomas Kuhn, but shared a number of common features with it. First, this new generation of thinkers opened the black box of technological developments. Instead of treating technological artifacts as givens, they analyzed their concrete development and formation, a process in which many different actors become implicated. In place of describing technology as autonomous, they brought to light the many social forces that act upon it. Second, just as the earlier, Kuhn-inspired philosophers of science refused to treat "science" as monolithic, but found that it needed to be broken up into many different sciences each of which needed to be independently analyzed, so the new philosophers of technology found the same had to be done with "technology." Third, just as the earlier philosophers of science found that they had to speak of the co-evolution of science and society, so the new, more empirically oriented philosophers of technology began to speak of the co-evolution of technology and society.

The new philosophers of technology saw, on the one hand, that the development of technology is accompanied by a transformation of society, but on the other hand that the process of technological development is determined by social-cultural factors. As technological artifacts are successfully introduced into society, society is transformed in unpredictable and irrevocable ways; but technological developments are given their form by cultural determinants. In the words of the American historian of technology David Noble, "Technological development, in short, is not an independent force impinging upon society from the outside, according to its own internal logic; rather, it is a social activity in itself, which cannot but reflect the particulars of its setting: the time, the place, the dreams and purposes, the relations between people" (Noble 1978, 248). Many different images and metaphors have been used to express this intertwining of technology and society in empirical studies of technology. Words like "technosociety" or "technoculture" are often used. Thomas P. Hughes speaks of a network of technological affairs that contains "a rich texture of technical matters, scientific laws, economic principles, political forces, and social concerns" (Hughes 1983, 1). Cockburn and Ormrod speak of "the mutual shaping of gender and technology," Schwarz and Thompson of technology that is "not some exogenous force—that has impacts on society," but rather "a social process that is extraordinarily inaccessible to us because we are so much a part of it." In opposition to classical philosophers of technology, these authors see the direction and outcome of this social process as in principle unforeseeable. They see, for instance, no inherent logic pushing technology and society relentlessly in the direction of greater efficiency or uniformity.

In general, the classical philosophical approach can be said to emphasize the "disembedding" processes of technology, in which cultural rootedness of traditional techniques is lost. This process of deracination no doubt formed and continues to form the conditions of possibility of the development of modern technology and of a technological culture. The classical philosophical approach failed to appreciate that another process is simultaneously at work, which involves the intertwining of technology and society and gives rise to a technological culture. Especially the idea of the autonomy of technology as Ellul develops it relieved classical philosophy of technology of responsibility for close philosophical reflection on this technological culture; the phrase "technological culture" for Ellul is even an oxymoron. His criteria for what is included in human culture excludes the possibility of a technological culture, except in the form of a monstrosity unique to world history (Ellul 1990, 175). Against this approach, I claim that it is precisely the task of an empirically oriented philosophy of technology to understand the co-evolution of technology and society in modern culture, rather than to evaluate it on the basis of a priori criteria.

The differences between classical philosophy of technology and the empirically oriented contemporary approach are evident not only in their approaches to culture but also in their approaches to nature. The classical approach takes it for granted that nature is "disenchanted" in the hands of the technological projects of modernity, cut up into however many ways or into however many pieces of matter that humanity finds useful. But as Simon Schama, for instance, reveals in his book on the history of the landscape, *Landscape and Memory*, perhaps in our approach to nature "we were never modern," to borrow Latour's well-known words. In the last centuries in the West, human beings have constructed many different, mythical forms of the landscape. All the different modern myths about forests and wilderness that are found in the United States, Germany, France, and England challenge the simplistic idea that the modern world conceives nature solely in terms of something useful for human beings, reducing it to what Heidegger calls a "standing reserve." The stark contrast between culture and nature employed by classical philosophy of technology, which partly derives from Bacon and Descartes, gives way to a more subtle interweaving of both.

Much of the experiencing of nature—not only within environmental and back-to-nature movements but also within contemporary environmental philosophy and philosophy of technology—that is hailed as a reaction to the so-called reductionist technological approach to reality paradoxically has arisen as a by-product of modern technological culture. Some recent studies, for instance, have discussed the experience of mountain climbing as a product of modernity, related to other experiences involving the search for and exceeding of limits—be they outer or inner—and the cultivation of new experiences and sensations. Mountain climbing presupposes not only a modern technological infrastructure, but also a very technical manner of locomotion using many kinds of high-technological devices, and its professionalization can only be understood as an accessory of technological culture itself. Much of what has just been said about the experience of nature applies in broad outline to environmental problems and the technologically mediated relation between nature and culture.

The issue here is to understand this new cultural constellation, rather than to reject it nostalgically in demanding a return to some prior, seemingly more harmonious and idyllic relation assumed to be possible between nature and culture. Such is precisely the approach of Don Ihde, who allows us at this point to see how technology, as a new and richer way of grasping nature, can give us a more expanded access to it.

This applies as well, in a more general sense, to all of the other thinkers discussed in this collection. In one way or another, they each have drawn from the empirical well. They stand in the middle of the world of designers and users of technology; they make abundant use of research into technology, especially from sociology and women's studies; and they communicate directly with technologists and engineers. All struggle with the question of how to come to terms with all this in a philosophical way. Even in cases in which their struggles are clearly still in progress, their efforts have produced a number of new and often surprising insights into contemporary technological culture.

BIBLIOGRAPHY

Arendt, Hannah. 1958. *The Human Condition.* Chicago: University of Chicago Press.
Cockburn, Cynthia, and Susan Ormrod. 1993. *Gender and Technology in the Making.* London: Sage.

Ellul, Jacques. 1988. 1990. *The Technological Bluff.* Translated by G. W. Bromiley. Grand Rapids, Mich.: Eerdmans.

Hughes, Thomas P. 1983. *Networks of Power: Electrification in Western Society 1880–1930.* Baltimore: Johns Hopkins University Press.

Illich, Ivan, and Barry Sanders. 1988. *ABC: The Alphabetization of the Popular Mind.* San Francisco: North Point Press.

Jonas, Hans. 1985. *Technik, Medizin, und Ethik.* Frankfurt am Main: Suhrkamp.

Mitcham, Carl. 1994. "Philosophy of Technology." In *A Guide to the Culture of Science, Technology, and Medicine,* edited by Paul Durbin. New York: Free Press.

Noble, David. 1978. "Social Choice in Machine Design: The Case of Automatically Controlled Machine Tools, and a Challenge for Labor." *Politics and Society* 8: 242–312.

Schama, Simon. 1995. *Landscape and Memory.* New York: Knopf.

Schwarz, Michiel, and Michael Thompson. 1990. *Divided We Stand: Redefining Politics, Technology, and Social Choice.* London: Harvester Wheatsheaf.

1. Albert Borgmann: Technology and the Character of Everyday Life

Pieter Tijmes

Of all those philosophers who concern themselves with technological culture, Albert Borgmann speaks with a distinctive and recognizable voice. His claim that contemporary culture bears the imprint of technology is not quite new, of course; predecessors include thinkers such as Hannah Arendt, Jacques Ellul, Arnold Gehlen, Martin Heidegger, and Lewis Mumford. But Borgmann articulates this vision in original ways. He discerns in technology an underlying design that he calls a *device paradigm,* with which he shows that the relation of human beings to reality is subject to change. While Borgmann finds this transformation regrettable, he offers a therapy at the same time as he makes the diagnosis—and precisely this is what distinguishes him from his illustrious predecessors.

Following a short biographical sketch, I shall discuss how Borgmann diagnoses the ills of contemporary life, what his concept of the device paradigm of technology is, and what its implications are. I shall discuss how he uses this idea in showing that technology is indeed a revolutionary factor in society, politics, and culture, and how he charts the course of these changes. What concerns him most are those changes in the human relation to reality brought about by the device character of technology. Borgmann does not hesitate to speak of the "loss of meaning" of things, and laments the transformation of culture at the hands of technology. The therapy he counsels involves a new engagement with things, one that does not involve turning away from technology.

• • •

Albert Borgmann was born in 1937 in Freiburg—"In the shadow of a Gothic cathedral, the Black Forest, and the university where Husserl and Heidegger taught," as he puts it at the beginning of a short autobiographical essay (1993). Those conversant with his work may already recognize traces of his heterodox Catholicism, love for nature, and respect for Heidegger. Concerning his relation to the latter, Borgmann has often spoken of his "respect, but never love for Heidegger," a phrase that indicates in a subtle way both his nearness to but also distance from that thinker. In 1958 Borgmann went to the United States, where he studied German literature. In 1961, after receiving an M.A. degree from the University of Illinois, he returned to Germany to study philosophy in Munich with Max Müller and Wolfgang Stegmüller. After completing his dissertation under Müller,[1] Borgmann returned again to the United States to seek a university position. He held several temporary positions before winding up in 1970 at the University of Montana in Missoula. In 1974 he published *The Philosophy of Language.* But in the mid-1970s he was forced to rethink his philosophical position. Heidegger had shown him the relevant philosophical problems, while political thinker John Rawls had provided him with the model of a solution. His next book, *Technology and the Character of Contemporary Life* (1984), reflected this new philosophical stance, though in his autobiographical article "Finding Philosophy" Borgmann claims to doubt whether in it he managed to weave both traditions together successfully. True or not, Borgmann has remained fascinated by both traditions, though as he began to develop his own lines of inquiry the references to other traditions became less relevant. Borgmann is often referred to, not inaccurately, as "an American of German origin"; still, from a European perspective, in many ways he is the most American of all the American philosophers of technology.

THE DIAGNOSIS OF CONTEMPORARY LIFE

Under the influence of technology, Borgmann claims, human beings become ever more detached from things. Things lose their weight and meaning for humans, slackening their bonds to us on many levels: physically, mentally, socially. Humans have but fleeting and often boring contacts with their surrounding material world. Borgmann couples this diagnosis with a suggested therapy; he seeks a renewed engagement *alongside* the usual, superficial dealings with reality. This cashes out to a plea for activities like running, gardening, cooking, and the culture of the table—practices that orient the lives of humans but coexist uneasily and with much tension alongside the prevailing activities of our fast-paced technologically driven existence. Borgmann does not want to condemn these latter activities, only to complement them. If the therapy does not sound intuitively pleasing at first, it is nevertheless important to understand his diagnosis. This will make his philosophy comprehensible and more easily evaluated; the diagnosis has its own qualities and can be analyzed separately.

Borgmann makes his vision of a positive relation with a reality due to technological devices plausible through his account of a *pattern* that he discerns in the use of technological devices. Technological devices, Borgmann says, deliver a product. A CD player delivers a product—music—in such a way that it is available to me at any hour of day or night. Each time that I use a CD player, it delivers the product with the same high quality. The CD, however, represents a high level of abstraction in the sense that I can't see from the disc itself whether the music I'm about to hear is made by a harmonica or an organ, by a soloist or an orchestra; I can't even tell whether a disc has pictures or music stored on it. Whatever kind of music I want I can find on a CD and play it in my living room.

Once upon a time, this took more doing. If one wanted to hear music, one had to stage an event that required an infrastructure and some organizational talent. Musicians were not always available to perform, the way they are on a CD. Only the very rich who were able to support court musicians could have music played for them whenever they wanted. The quality of the performance could not be guaranteed in advance: the musical event could be a good day or a bad day. We are more blessed in that respect than the Margrave of Brandenburg, Christian Ludwig, though in comparison with the way he heard the Brandenburg Concertos, the way we hear them in our living rooms lacks something in the way of spectacle and atmosphere. Our CDs, too, of course, require organizational talent and technical infrastructure, but in a much different way than those earlier musical events.

Modern technology is all about product, while the device itself withdraws. The consumer knows nothing of the science of sound recording and reproduction, nor of the manufacturing processes involved, nor of the distribution system that brought the CD from Japan (or wherever) into the home. Modern technology shows the device responsible for the product but hides the whole "machinery" making it possible from the notice of the consumer. The products are destined for the consumer; the technology is the business of the engineer. It hardly needs to be said that the advantage and disadvantage of technologically available recorded music cannot be weighed in an easy and proper way: the advantage (availability) and disadvantage (lack of engagement) are heterogeneous quantities. The valuation that expresses itself in enthusiasm or disapproval is tightly bound to different effects.

THE TECHNOLOGICAL PATTERN

Borgmann seeks to map out technologized society—what Heidegger designates with the term *Gestell,* or framework—with the aid of his device paradigm. This method has a great advantage over Heidegger's own method, for instance. According to Heidegger, the modern technological approach to reality delivers over to humans a universe of manipulation in which they express their will to power. If one asks where this will to power comes from, Heidegger's answer is that this specific modern approach to reality is not a human choice; it belongs to the fate of a specific period, it is the "sending of being" in the modern age. In what follows I shall show, among other things, how Borgmann's characterization of technology can help us understand how attractive technology has become in our society, and why.

Modern technology is modeled on a specific pattern that we can recognize in all of its forms even outside the practice of engineering. That pattern consists of a device that makes commodities or products available on demand. One important characteristic of such a device is a sharp separation between means and end. The end is the product, while the means are exclusively in the service of this product; these form, as it were, the hidden machinery. A watch has the function of providing the time, and its machinery has only that one end and cannot be used for anything else. Discarded watches are useless and cannot take on another function. The same holds for discarded washing machines or dryers, which are good only for the scrap heap. However, from the outside of a watch one cannot directly see whether its machinery is digital or analog, meaning that its machinery can be radically altered without affecting the function. This is what Borgmann has in mind when he says that the means and end are sharply separated from each other. The machinery has been set up to produce the end in a safe, easy, instantaneous, and ubiquitous way. It is a pure means, and as such withdraws from sight.

For this reason, one might say that our existence is characterized by the permanent presence of technology that does its work in secret. Clear water flows conveniently and safely from the faucet whenever we want, and we use it without dwelling on the hidden conditions of this comfort. We trust in the availability of the water and anticipate that it will always flow when we open the faucet. *That* the faucet works is a matter, not for us consumers, but for the engineers, the technicians, the experts. The same holds true for all of the devices that surround us in our everyday lives—automobiles, vacuum cleaners, washing machines. Whenever these break, we take them to the experts to be repaired, or junk them if they cost too much to fix. We have internalized, as it were, the division that the device effects between pure goal (its product) and pure means (the machinery that gives rise to the product) as a specific modern posture with respect to reality—an internalization that represents an extreme form of functionalization. We like the product and count ourselves lucky that we don't need to concern ourselves with the conditions that have to be satisfied to produce it.

This pattern of the device paradigm is not limited to technology itself but also occurs in adjacent areas. Technology, one might say, institutionalizes this pattern and its progressive advancement—even in our thinking. The device paradigm is seamlessly incorporated into consumer society, apparently realizing the dream of effortless consumption. Microwaves can make a store-bought meal ready in five minutes, and shorten even the lengthy time it used to take to prepare a big holiday meal in the old-fashioned, tasty way. Stop & Shop offers not only all the ingredients to make traditional meals, but also fully prepared pizzas and pre-cooked chicken and ribs. One might think of the supermarket here as the device that delivers "buy and serve" products, but in these products we can also recognize the devices that produced the old-fashioned Thanksgiving or Christmas meals. The holiday rituals surrounding the preparation of such meals have been replaced by the trip to the supermarket. We sacrifice these holiday rituals for convenience, and this convenience is linked with less engagement.

THE PAST AS CONTRAST?

The revolutionary character of the device paradigm becomes clear when contrasted with the role of technology in traditional society; the contrast serves to rob, as it were, the paradigm of its obviousness. In modern society, goods are produced in order to enrich our lives and make them more pleasant; in creating its stream of goods, technology thereby plays the role of lightening our burdens. Heating is now technologically available everywhere, in all nooks and crannies of every house, office, vehicle, and so forth. This was not always so: in earlier times not every room in the house was heated, no room was heated all the time, and there wasn't much difference between the temperature inside and outside a vehicle like a carriage. The production of warmth, to put it in modern terms, was based on a division of labor in which the entire family took part. The open fireplace generally would be cleaned and reset, and thereafter continually restoked, by the woman of the house early in the morning.

Procuring the wood was the job of the man of the house, while the children were responsible for making sure that there was always enough wood available in the bin. Everyone in the family, in short, had a role in heating the home. One finds another example of social division of labor in the obtaining of water from a communal well. The well not only fulfilled the technological role of providing the water supply, but also and at the same time served as a social meeting point, where one could pick up all the village gossip. Abraham's servant found Rebecca, the future wife of Isaac, by a well, while a well was also the site of the first kiss between Jacob and *his* future wife Rachel.

Following Heidegger, Borgmann clarifies the difference between traditional and modern technology via the pair of concepts "thing" and "device"; devices are contextless technological means, while things, by contrast, are socially embedded. The term "contextless" is not pretty, but the meaning is clear. Water doesn't flow from every spigot; whether it does so or not is a function of its hidden machinery, which is looked after by the engineer. This hidden machinery, rather than the social environment, provides the context for the spigot.

By contrasting the spigot (a device) and the well (a thing), we can read off interesting differences between modern and traditional technology. A well requires more labor and exertion than the spigot; the human-reality engagement or interaction is greater. Whoever draws water from a well must be familiar with its machinery: the steadiness of the pulley, the wear on the rope, the way the water table varies with the seasons, and so on. These features are visible and accessible for the required interventions by the users. The reliability of the water supply is the concern of all those who draw the water. When Borgmann asserts that things have become *devices*, this is not merely a matter of a technological improvement but involves, more importantly, a change in a complex social process.

The love that Borgmann lavishes on his descriptions of the interactions and social embeddedness of traditional technology leaves him open to the charge of nostalgia. This charge is not necessarily interesting; what's wrong with a little nostalgia? It would be interesting only if it revealed weaknesses in his account of the difference between present and past. But it is difficult to see any. The weekly ritual of stocking the woodpile next to the fireplace in days of yore has a completely different meaning than today. In the first case it concerns a life necessity—even survival—and in the second case a relaxing exertion in preparing a convivial atmosphere for gatherings of friends and acquaintances. Seemingly similar activities are in reality extremely different. Completely lacking from the latter happy social occasions is the labor, responsibility, and urgency attaching to the former.

Still, the positive virtues of the technologically framed world should not obstruct our vision of its drawbacks. One could think of the reduced physical contact with reality as well as the social atrophy brought about by many forms of modern technology. On the other side, there is a certain irony in the thought that the attempt to adjudicate past versus present by *measuring* certain isolated values is itself a technological process more befitting the present than the past. It is therefore not a simple matter to evaluate the changes resulting from modern technology. It is hardly meaningful to pose, let alone answer, the question, "Have the changes effected by modern technology led to good or bad?"

In short, we face two key problems in attempting to contrast the present with the past. First, the replacement of the thing by the device has altered the conditions by which we could make a good comparison; such a comparison would be legitimate only if the conditions remained the same. The evaluation depends on the parameters that we choose for the purpose of making the comparison. Both assertions—that modern technology means progress, and that it represents a desiccation of social life—can be defended on solid grounds. Second, modern technological (device-oriented) thinking reveals itself at work, though concealed, in the very attempt to compare thing and device, though as we have already seen the thing cannot be thought of in this way.

THE DESIRABLE FRUITS OF MODERN TECHNOLOGY

One prominent feature of modern life, Borgmann points out, is the sharp division between work and free time, and he sees this as connected above all with the division between machinery and products in the framework of modern technology. Free time consists in the unencumbered enjoyment of consumable products, while work is generally oriented to the construction and maintenance of the machinery that sees to these goods. The attraction of technological progress lies in the lightening of heavy work loads, better health, possibilities for cultural development, and free and unlimited consumption of goods. We prefer to enjoy the blessings of technology in the form of an increase of goods for consumption. The increase in productivity therefore translates not into more free time but into expanded consumption. In the contemporary world, Western society plucks the fruits of technology in the form of mass consumption.

Already at the beginning of the seventeenth century, Francis Bacon, in his book *The New Atlantis,* expressed a sunny belief in the liberating and enriching lives that would be bequeathed us by science and technology. It took some time for technological society to get under way, but when the industrial revolution finally arrived in the second half of the eighteenth century, it brought along exploitative labor practices, bitter misery, and utter poverty. It would take still more time for the promised predictions about technology to come true—and it is also understandable that technology initially met with opposition and even aggression (for instance, by the Luddites; see chapter 6 of this book). Only at the end of the nineteenth century and then finally in the twentieth did technology lead to widespread prosperity in western Europe and North America. The history of technology and the spreading of prosperity is thus a complicated story: initial exploitation, misery and poverty, followed later by plentiful consumption. Borgmann, however, focuses his interest on technology as it characterizes our *contemporary* life. According to him, our entire reality, including nature, culture, and social relations, is modeled in accordance with the device paradigm. His working hypothesis is that we judge our lives and our society in the light of technology. Before returning to Borgmann's interpretation of the workings of contemporary technology, it will be enlightening to follow Borgmann's train of thought about the political implications of the device paradigm in liberal-democratic society.

POLITICS AND TECHNOLOGY

Ideas like freedom, equality, and self-realization are central in today's liberal democracies, and are even mutually supportive. As free-standing ideals they tend to have short lifetimes; if one is absolutized, it is at the cost of the rest. In this sense they need each other: freedom by itself implies an assault on equality, the ideal of pure equality is an obstacle to freedom, and self-realization can scarcely arise as an ideal without the counterbalance of equality and freedom. One part of the political ideal of a free democratic society is that the individuals in it strive freely, and as equals, to realize their respective abilities. The term "self-realization" is formal in the sense that it is not prescribed in advance which abilities these are or in which direction they are taken. This triad of realization, freedom, and equality stands in a strained relation with the principle of the capitalistic market as the regulator of opportunities and goods, which liberal democracies also hold sacred. A difference of opinion presides over this strained relation between political ideals and the market. To some, the market provides a very efficient—in a certain sense unimpeachable—way of adjudicating positions and opportunities and of correcting gross inequalities; others, however, do not share this optimism regarding the market's functioning.

Current political debates are dominated by financial and social-economic issues. These debates proceed as if economic politics provided the boundary conditions for the resolution of all other issues, including ecology, education, and social security. The logic that every dollar can be spent only once governs acceptance of the result; the quarreling takes place mostly about the means. Discussions about the good life, the presumable goal that the means hope to realize, rarely arise.

In this context one may argue the case for affinities between democracy and technology. Democracies, as previously mentioned, do not specify the form of the good life, nor the lifestyles of their citizens. They try not to be paternalistic, leaving the responsibility for how to live up to the citizens themselves. Technology seems to hook up with this extremely well; it promises to open up new opportunities for self-realization, liberating us from life's chores, while continuing to leave open and indeterminate the question of the good life.

But there is more than just an affinity between democracy and technology; democracy also *depends on* technology to realize its goals. Many practical obstacles stand between the democratic ideals of equality, freedom, and the good life and their realization. How can a democracy cope with concrete inequalities among its citizens? Striking inequalities of material, social, and cultural opportunities are best dealt with, not by redistribution, but by increased productivity. This role is perfectly suited to technology, through increasing the stream of goods. The good life thereby receives a concrete content along the lines of technology. Although technology is conceived instrumentally, it makes the good life equal to consumerism—which undermines the claim for the value neutrality of technology.

Technological products such as the automobile, telephone, and television have changed the world so much that a return to a world without them is no longer even thinkable. Choices made about technology cannot be revoked. A decision about the content of a good life is implicitly made once these devices are accepted. While choices between specific products might well be made impartially on the basis of social and economic structures, a choice has already been made in the conflict between an engaged dealing with things versus a consumerism of commodities. To put it another way: technology is never offered to us as a way of life that we can prefer above other ways of life, but is advanced as a basis for choices. Humans can make choices as consumers, but technology itself is always present as the basis for the choices. According to Borgmann, a choice against technology, or rather against a technologically impregnated democracy, would amount to a choice against freedom and for prejudice and paternalism. To reject the promise of liberation and enrichment is to accept oppression, poverty, and suffering—not, therefore, a real choice. Technology is deeply anchored in our existence yet tends to become invisible. Put more strongly: to the extent that technology withdraws from sight, it becomes that much more astounding.

In short, the promise of technology develops into a lifestyle in which the political ideals of freedom, equality, and so on undergo erosion; conversely, the promise of democracy is technologically specified. In this context, politics can take on mainly the character of a meta-device of technological society. Even political behavior is modeled on the device paradigm. Whenever a crisis occurs, a special commission is appointed to study what broke down. The role assigned to the citizen is to be a supporter of the government machinery by, among other things, paying taxes. The government minds the technological store—and in the shadow of this task political party identifications hardly matter.[2]

Modern technology provides the inescapable horizon of our existence. In a sense, we are all accomplices to its domination by virtue of the fact that our enjoyments strengthen its hold. If despite this we ever dare to express doubts about this hold, we are led to reflect on norms and values. But this reflection is like running in loose sand, and jumping is out of the question. It is of course important which norms, values, and goals are selected, but within the technological framework they remain commodities to consume. The political discussion always comes down to a matter of money or its distribution: political goals are discussed in market terms. However deeply they purport to cut, proposals for greater responsibility, greater technical efficiency, and more citizen participation leave the basic structure of technology untouched. These reform efforts may arise from a dissatisfaction with technology but, however strong, are doomed to fail because they remain on the side of technology, and are blind to its idiosyncrasies. This discussion is quite different from the idea of the so-called *technological fix*, in which technology is pitted against technology.

HAS BORGMANN BURNED HIS BRIDGES?

It might appear as though many possibilities for a normative discourse about how to lead a good life have gone up in smoke. The very fact that Borgmann is so persuasive in his demonstrations of the revolutionary power of the device paradigm leaves him with little normative wiggle room, because things evidently are regulated in the shadow of the dominant paradigm. Technology preordains or at least channels our decisions. However, Borgmann does not share the liberal view that the good life remains a still-to-be-specified aspect of our society. Again, this is not to deny that all sorts of normative questions crop up in certain sectors of technological society, such as the honesty, integrity, and responsibility of those functionaries who tend to the maintenance and expansion of technology; or litmus issues of the Christian right, such as abortion, euthanasia, pornography, and capital punishment. The all-important issue of technology and its goal of endless consumption never comes up for discussion.

After burning his bridges, how is it possible for Borgmann to reopen the question of the good life? This is a tricky one, for Borgmann, who does not yield an inch here to postmodernists, denounces the practice of many philosophers who reach strong conclusions on the basis of weak assumptions. They begin with certain specific axioms that are self-evident or noncontroversial, and wind up with a vision of a well-ordered society or the existence of God—rich conclusions. John Stuart Mill, for instance, invites us to strive to seek to do nothing less than the greatest good for the greatest number of people, while John Rawls would have us adopt the perspective of a rational person striving after his own interest among equals. According to each of them, rational calculation shows that the result will be an almost ideal society. These approaches are modeled on a certain view of philosophical procedure. The trouble is that if one begins with few assumptions, one ends up with few important conclusions; one can arrive at important conclusions only by first assuming a lot. The philosopher's secret, then, is often the ability to smuggle in extra suppositions unnoticed along the way. Borgmann refuses to operate in this manner, convinced as he is of philosophy's inability to press toward important truths via an internal route.

In place of this traditional, quasi-rational discourse, Borgmann introduces what he calls a "deictic" discourse, from the Greek for "to show" or "point out." He describes this as a discourse that is preeminently "testimonial" or "appellative." Deictic discourse does not lead to a logical deduction or scientific conclusion but rather to a plea for assent. It points to a new engagement with things and respects the other's integrity and feelings. It has been said that Borgmann's philosophy is in the service of things—things not in the sense of pebbles and flowerpots, but of what Borgmann calls "focal things." The term "thing" is the counterpart of "device." The term "focal" refers to the central meaning that a thing can have for our lives. The Latin word *focus* means hearth. In a pretechnological house, the hearth represents the center of light, heat, and daily activities. For the Romans it was the place where the house gods dwelt, where they ate and gave offerings to the house gods, and where marriage ceremonies were performed. In Greece a baby was accepted as part of the family and household when it was carried around the hearth. A trace of these old traditions is carried over in the central place that an open hearth usually occupies in modern houses. Above and below the mantelpiece one no longer hangs pictures of the gods, but photographs of loved ones and family members who have passed on. It is also a place where clocks are often kept. In short, the word "focal" refers to the granting of a central orientation.

Focal things incorporate a sense or meaning. Their sense or meaning cannot be derived philosophically, but may be brought out via more literary means. Deictic discourse does not negotiate commitments, but rather exhibits the importance of a thing or practice. It is an enthusiastic discourse, and the enthusiasm is contagious. It teaches us what we ought to do by telling us what the activity is. This deictic discourse, Borgmann claims, is a way of bridging the gap between "is" and "ought." Deictic discourse is contestable; even though it points away from itself toward things (or to practices associated with things), it always remains possible that the meaning of the thing is not brought out eloquently and in its own right.

One searches in vain for a definition of focal things; rather, Borgmann shows them off deictically—testimonially or appellatively. Focal things are of different origins and often fulfill different roles, making it impossible to point to a specific definition. Focal things form a group and belong to each other without sharing a fixed general characteristic. They exhibit a family relationship, so to speak. A further complication is that Borgmann also discusses the *practices* that are deployed about things, and therefore focal practices—those practices that center, orient, and enlighten our lives.

Borgmann discusses a number of things and practices belonging to this privileged "focal" family. The earlier examples are the clearest and most appealing to us; the modern examples are more vulnerable and ambiguous. Thus Borgmann speaks about temples and cathedrals (things), but also of processions and celebrations (practices) central to the meaning and coherence of life. The cathedral of the Middle Ages is a point of convergence of its lifeworld, embodying the medieval vision of the world: the beginning, middle, and end of the history of salvation, the hierarchical order of reality that culminates in God, and encompasses the place occupied by each individual. To be sure, there is no present-day equivalent of the medieval cathedral—not even the technology laboratory, where we send our cleverest young men and women. Religious rituals such as attending church, prayer, and Bible readings are focal. In certain respects these things serve as a good contrast to the world of devices, in which everything is available on call. But examples taken from religious activities can mislead by suggesting a stronger kinship between the focal and the religious than Borgmann intends. In his appeals for focal things and practices against the dominant technology, Borgmann in no way makes a case for Catholicism. Practices such as running, reading books, dinner table conversation, and gardening count as members of Borgmann's focal family. Such practices, he stipulates in defending their inclusion, do not have to be oriented to specific ends or goods but can be done as ends in themselves. If they are carried out with specific goals in mind, of course, they cease to be focal activities and must be interpreted instead as technologies in the service of an end. Some examples of specific focal things and practices that forsake the device paradigm will help clarify what Borgmann means.

FOCAL THINGS AND PRACTICES

Nature and culture are hardly independent things. In Europe one speaks of cultivated nature—a nature of fields, meadows, and woods. On the American continent it is possible to come across nature unencumbered by cultural contributions: "wilderness." Wilderness, Borgmann claims, possesses focal power, and is a focal thing. What does that mean? Wilderness preceded human history and owes nothing of its beauty to humanity. Borgmann calls nature in such an untouched state a challenge to technology, because it represents the contrary of technology. This is not to say that it is something to be subdued, but rather that it is capable of summoning human beings to a special kind of experience. It is capable of eliciting a respect that arises out of something other than enlightened self-interest or the valorization of sustainability, a respect that arises from the recognition of something other in its own right. The technological universe, however, is inhospitable to language that would express and affirm such a recognition, and constantly threatens to undermine the language of deictic reflections on wilderness. That something comes to be expressed at all is due to the power of nature itself; nothing else is as eloquent.

This assertion must not be confused with the claim that nature possesses a value, based on the supposition that the preservation of nature requires a justification. Thus wilderness is sometimes defended in the name of formal properties like complexity or diversity. Borgmann is dubious of this strategy, as he is of values such as beauty, stability, and integrity, which are suspect in deictic discourse. What he sees at issue here is not judging nature in the light of beauty, but the reverse: learning from nature what beauty really is. Terms such as "stability" and "integrity" are so flexible that they can be used to judge anything. Nature itself, in other words, begins to emerge only when such concepts are given up, as in literature and poetry. Only then can we learn what it is to be confronted with something that is other, and greater, than ourselves, letting it stand in its own glory without having it serve our own interest. Wilderness, as it were, can speak to us in a pre-technological language. In experiencing wilderness, we experience ourselves anew. With respect to wilderness, we are neither consumers nor conquerors; those who encounter it are summoned with their entire being. What we get as consumers from nature is limited: water, firewood, berries, and fish perhaps. But the blessings of technology also feed us and keep us warm. Without technology, surviving in the wilderness is difficult and risky. Wilderness therefore teaches us to cherish technology, while thanks to technology we can respect nature.

The point is not for philosophy in its deictic deliberations to cite poetry, but to reflect on conditions that hide and disclose matters of ultimate concern. Borgmann learns from nature that pre-technological things are not simply hopeless relics of an unrecoverable order, but beacons scattered here and there in the midst of technology. Wilderness is indeed focal, because it can illuminate for us a new orientation toward technology.

Other focal things and practices include music making, gardening, and the culture of the table. Borgmann calls these things and practices "inconspicuous." They flourish in the margins of public attention, and form no unity. But Borgmann associates them with the idea of a reform of technology. By that, he does not have in mind the implementation of a new master plan for the universe of technology, but rather the discovery of sources of power that do not deny or rival the dominant forms of technology so much as limit or complement them.

In the discussion of the culture of the table, Borgmann refers to the simple joys of washing salad leaves and slicing bread—focal but also highly vulnerable practices. The discussion clarifies the aim of his reforms: to accept technology and simultaneously recover certain focal daily practices that have been eroded in American life by the technological order. Thus Borgmann describes running as the opposite of the device paradigm. "The long-distance runner is the least of all athletes. His sport the least of all sports," Borgmann quotes marathoner George Sheehan as saying (Borgmann 1984, 202). Running consists simply of moving oneself through space and time, step by step. But there is splendor in this simplicity. Borgmann describes eloquently how effort and joy are one in the runner, overcoming the split between means and ends, labor and leisure. This unity between achievement and enjoyment, competence and consumption is only one aspect of a central wholeness that running provides us. Running engages mind and body. "The mind is the sensitivity and the endurance of the body" (203). Running in its wholeness, Borgmann says in his remarks on Sheehan, is something utterly different than exercise for physical health. Borgmann cites observations by marathoner George Wood that he feels more in touch with the realities of the inner city while running through it slowly enough to take in the grim details, and that he experienced such a great proximity to the crowd that he felt he had to run well *for* them, in response to their support. Sheehan comments that, for him, running discloses the divine; when he runs, he "wrestles with God" (204). Serious running brings us to the limits of our existence. We meet unbearable pain. Sometimes this experience obstructs our vanity and ambition. But running can also bring us nearer to the point where suffering allows us to experience our greatness. Thus the experience of these marathoners provides us with a possible place from which to escape from technology, metaphysics, and the gods of the philosophers and to reach for something more like the personal god of Abraham, Isaac, and Jacob. Borgmann's discussion of marathoners is yet another instance in which he seeks to bring an experience deictically into language.

REFORMING TECHNOLOGY

At the heart of the reform of technology lies the attempt to make life hospitable, within the technological universe, for focal things and practices. "Reform," so conceived, consists not so much of a change or repudiation of the technological paradigm, but of an attempt to circumscribe it so that focal things and practices receive a central place. The device paradigm is put where it belongs—and it belongs in the background and on the periphery of focality. When this happens, the device paradigm ceases to determine the characteristics and dominant manner of dealing with reality. Focality is the antidote to the sway of the device paradigm over our lives. This antidote is neither pre-technological nor anti-technological, but meta-technological; it opens the door to the meaning of things and to human dignity, it tells human beings what they really need in the world, it supplies us with riches—not of course to be confused with the technological superabundance of material goods—it summons us to care for the safety and well-being of human beings and the environment. According to Borgmann, there is much intuitive evidence for the thesis that focal devotion points to an intelligent restriction of technology, and invites us to toss a significant amount of technological junk out of our lives.

The entire technological complex, therefore—devices, machinery, and commodities—are to be arrayed about focal practices. The idea of the good life then receives its content from focal participation: the good life is focal praxis. But the good life cannot be attained unless it unfolds both in the world of work and in the public domain. It remains a makeshift solution, but an individual can make a beginning on the personal level by centering one's life around focal practices. Borgmann emphasizes that, in our time at least, focality does not have a center but only a disparate character.

Given Borgmann's noble words about the reform of technology and its permeation of the world of the individual, one naturally grows curious about what he has to say about reformation on the collective scale. How can the reformation be spread to the workplace and the public world? Better: how can we transplant focal concerns to the political sphere and to the machinery of technology? It should already be clear that the public world cannot be structured around one or another focal gestalt, because focality has a disparate and heterogeneous character. The reforming of technology does not consist of a definite, implementable plan, but rather of the flowering of focality. This means, however, that human beings must be prepared to pay an economic cost, for the turn to focality bends away from technology. Studies show, Borgmann claims, that human beings are ready to take this step. In this framework a collective affirmation of a reforming of technology is necessary, and must emerge from a public deictic discourse, one in which the difference between quality of life and standard of living is sharply articulated. The quality of life has to do with goods that cannot be commercially produced or individually procured. The technological universe, therefore, will be recast differently.

Borgmann finds support for these ideas in the notion of a dual economy advocated by certain economists. Such an economy consists of two sectors, one comprising large corporations and the other smaller businesses. Borgmann distinguishes between the two in that the former is almost completely automated while the latter are local and labor intensive. In Borgmann's view, the reforming of technology requires that we recognize, affirm, and promote this "dual economy." He appeals to Galbraith in support of this view; Galbraith claimed that such an economy already exists, though he draws the difference between its two elements somewhat differently. Borgmann refers to Mumford as well, who distinguished between authoritative and democratic technology; Mumford sees the first as systems oriented, very powerful, but inherently unstable, and the latter as technology oriented to human beings, relatively weak, but resilient and durable. Borgmann is extremely wary of accepting any of these distinctions; in referring to them he is mainly interested in the possibility of an engaged work that contrasts with technological work. The issue of reform on a collective scale is developed in a later book in another way.

REALITY, CELEBRATION, AND RELIGION

In 1992, Borgmann published a book entitled *Crossing the Postmodern Divide*, which essentially carries forward the insights of his earlier book in the form of a critique of culture. In it, he continues to write and think in terms of the device paradigm, though he no longer uses the term. In this book Borgmann outlines a project of modernization along the lines that he had worked out in the context of technology and economy. This project has fallen out of favor among postmodernists and others because the remorseless realization of modern ideals has turned against the project itself. The domination of nature has turned out to be problematic; the sovereignty of the individual and equality of opportunities have not been realized. Postmodernism found the modernist project to be a deception and turned against it, but Borgmann finds that what postmodernism has to offer in its stead is incomplete. Borgmann has his own alternative, which he calls *postmodern realism.* Three of its key points—realism, communal celebration, and religion—can be thought of as pieces of a puzzle that were missing from his first book.

1. Realism

In *Crossing the Postmodern Divide,* Borgmann again speaks of focal things as eloquent reality, to be contrasted with the (hyperreal) products that are made available on demand by modernist technological machinery. The concept of the real has always been battered by science and technology, but never as much as in our age of multimedia communication and virtual reality. Which eloquent reality does Borgmann have in mind? His reply: the reality that lies between the microworld of atoms, subatomic particles, and their constituents on the one hand, and the macroworld of suns and galaxies on the other. But even this reality is not unambiguous, whence his need to distinguish between the real and the hyperreal.

Borgmann sees an underlying coherence throughout all reality that runs from wilderness—the clearest voice of eloquent things—through animals and plants to the matter such as stones, clay, wood, and fiber with which human beings come into intimate and active contact in manual labor. "Nature" does not disappear in the crafted piece, but becomes observed and celebrated in new kinds of skills. Things can be focal in the care of human practices: a wilderness in a hike, a horse in grooming and riding, a fishing rod in fishing. In one and the same breath, Borgmann now broadens the family of focal practices from the culture of the table, reading poetry, and worship; then adds in passing that this last is more than just one of many focal points. All focal practices, he argues, display an analogy with worship as a concentrated way of dealing with reality; evidently, Borgmann holds the religious attitude toward reality as a model for focal practices, and it is no coincidence that he borrows the phrase "the ultimate concern" from theologian Paul Tillich.

2. Communal Celebration

It should be obvious that Borgmann is not infected by that great phobia of liberal thought: namely, specifying the character of the good life on the collective level and intervening in society to achieve it. He is wary of modernist successes due to individualization and the atomization of society. He does not think that much can be gained from the justice and solidarity that many say they strive for, because liberalism rejects so dogmatically the idea of community. Liberal justice, he says, is dangerously incomplete. But neither does Borgmann belong in the camp of republican champions of the cause of communitarianism. Their communitarianism, he finds, is undermined by their dependence on the dominant technology and economy. (Hyper)modernism is always the enemy of genuine urbanity; Borgmann in this connection refers to the automobile as the catalyst of urban destruction.

At this point Borgmann steps forward as an ambassador of sports, art, and religion, for he maintains that the realization of public life cannot take place in a hyperactive glorification of the everyday (work work work!) but rather in festive celebrations of the sort provided by these activities. He invites us to his home town of Missoula, Montana, to attend a baseball game, which he describes, using lyrical language, in terms of the essential features of celebration: reality, community, and divinity. The reality of the game consists of the flat field, the green lawn, the bat and the ball. Borgmann celebrates this reality and says it should not be impoverished by a hyperreal environment: Astroturf and a domed, air-conditioned enclosure. The definition of community of players is anchored in the game. They pretend not to share their possessions and convictions. The rules of the game and the boundaries of the field mark the limits of their teamwork. The game itself is shared by the spectators and makes of them a community as well. When reality and community work together in this way, then divinity descends upon the game, a divinity of an impersonal but yet potent character. Such celebratory events do not take place in any quarter of the technological universe; technology indeed constitutes a threat to such celebrations. When modern technology moves in on such celebratory events and attempts to take over, it undermines the key elements and can eventually obliterate them completely—as it often does, for instance, in pop concerts. The game becomes a commodity, something for the spectators to consume.

The task of postmodern realism, Borgmann concludes after reviewing artistic events in Missoula, is the care of communal celebrations. Political steps must be taken to make a large and central space for art and athletics in our local communities. Opportunities for uniting reality, community, and divinity in celebration must be supported. In the absence of political action, such opportunities will remain marginal in the social environment. Why all these communal celebrations? Borgmann expects much from the contagious enthusiasm of these celebrations; he sees them as coming to animate society and functioning as a panacea for the evils of hypermodernism.

Borgmann concludes his book with a discussion of a religious celebration that highlights for him a crucial fact of celebration. Human beings ask after the beginning and end of all things, he asserts, and situate themselves by their answers. Sports and art have much to offer here, but not everything. He then sets out for upper Manhattan, to the Cathedral of St. John the Divine, a congregation open to all classes, tongues, and denominations, where one is surrounded by the poor and homeless. Again the coordinating elements of this ritual are reality, community, and God. It is remarkable that Borgmann draws our attention to such events in the context of his philosophy of technology, but he makes clear why.

3. Role of Religion

Borgmann's view of the role of religion in other activities has already been touched upon. In an essay entitled "Technology and the Crisis of Contemporary Culture" (1996), he provides a new summary of his thinking as well as a justification for his Catholicism. In it he speaks of three spheres: that of the device paradigm, that of the sacred, and that of the Christian religion. He argues that the culture of modern technology and (Christian) religion are mutually exclusive in several respects. The sphere in which grace and sacrament have a role is quite different from the one in which commodities are made available on demand. A technological culture is inhospitable to grace and sacrament: it welcomes and is fully prepared for the notion that human beings are incomplete and deficient, and sets to work making a product to remedy the lack. As a result, where the device paradigm holds sway, the formative power of religion is weak. Thus it is our duty as Catholics, he says, to strengthen reverence and piety wherever we can find them. He considers his allies any environmental thinkers, artists, or scientists in whose work he discerns a respect and wonder that he himself articulates in religious terms. He considers public, joyous celebrations to be signs of the holy, and sporting events or artistic performances that are experienced as unanticipated and undeserved gifts to be signs of grace. Thus Borgmann wants to walk from the secular culture through the sacred and divine to the Christian space of sacraments, and back again. One would think that this is a project in and of itself; in any case, we know its basic blueprint.

INFORMATION THEORY

In 1999 *Holding on to Reality* appeared, Borgmann's most artfully composed book to date, in which he considers technology yet again, this time from the point of view of information theory. This perspective is elaborated in a crystal clear manner via a careful architecture. It is unmistakably in the tradition of his other masterwork, *Technology and the Character of Contemporary Life*, which makes forays into politics and economics. These are extremely interesting and call for further illumination and development, but in Borgmann's universe these now appear to be of minor importance. His premier importance lies in the characterization of technology and its consequences for the thinking and behavior of the average person—for you and me. Borgmann diligently pursues his own way in this new book toward a cultural interpretation of information technology. From the beginning, the political implications of this revolutionary development have been shoved to the side, because "it tends to divert us from the deeper question of whether the recent and imminent flood of information is good for anybody, rich or poor" (Borgmann 1999, 4). His opinion of this development is unambiguous: information technology is the high point of the semantic ice age, as he characterizes modernity. This harshly critical position is sometimes softened by conciliatory remarks about the need to seek a certain balance between the different kinds of information with which we are besieged, a balance between modern *technological* information on the one hand and the older *natural* and *cultural* information on the other. His slogan, "holding on to reality," does not make this task easy, as we shall see. The result, however, is an interesting theory involving the embedding of information in a spirituality with a heavy deictic cast.

Borgmann speaks about three kinds of information. The oldest is *natural* information, which contains information about reality. He finds this information already in the late and final stage of human evolution. The hunting and gathering cultures give us the earliest complete picture of the human condition. This brings him to the next stage: The ancestral environment is the ground state of information and reality. Human beings evolved in it, and so did their ability to read its signs. It is reasonable, Borgmann says, to assume that the attunement of humans to their original environment felt good. Here we find the basic and deeply pleasant structure of information. The ancestral environment was profoundly coherent because of the regular interplay of signs and things. With loving care Borgmann describes how natural signs disclose the environment. Some natural signs stood out as landmarks from among the conspicuous and transitory signs of creeks, rocks, trees, and tracks. Landmarks were focal points of an encompassing order. To pass beyond one's landmark was to lose one's orientation. Reality expressed itself with great eloquence. The meaning of a sign is not something selected or projected by human beings, but sent out by reality, though the human recipients must have the ability to glean the message from the sign.

In this formulation one feels Borgmann's phenomenological heartbeat. When signs begin to speak, they acquire personality. He refers to that moment in Genesis when eloquence rises to its highest in divinity and its most personal in the voice of God. The commanding presence of the divine voice suddenly spoke to Abraham and summoned him from the civilization of Babylon to a kind of environment and life that was closest to the ancestral condition of humanity. When, by contrast, *science* begins to describe reality, it has an eye only for general, law-governed structures. This gives rise to a kind of information gap between structural information that is uncovered by science and the contingent information about the expressive faces and eloquent voice of people and things. In modernity, the meaning of reality has declined and become occluded. It has been reduced to contingency—the unexplainable residue of accident and randomness. This is not true in an absolute sense, for Borgmann also uncovers indications of scientists "attesting to faint echoes of divinity, to the gratuitous beauty and sacred magnificence of reality" (33). It is clear that Borgmann highly values natural information in its eloquence. This openness to a speaking reality is also indicated in the title of his book: *Holding on to Reality*. In a plea for "focal things and practices" that is already familiar from his other books, Borgmann wants to provide something to help us winter over in this semantic ice age.

The *cultural* information—the second type—does not concern itself with natural signs like smoke and clouds, stones and footprints, but rather with conventional signs: letters and texts, lines and graphs, notes and scores. When cultural information steps into the world, it is not exactly as welcome as roses in May. Compared with the riches of spoken language, writing amounts to a radical abstraction, an idea encapsulated in the difference between, as we say, "the letter and the spirit." Borgmann recalls Plato's belief that philosophy is possible only in the form of a living interaction between people. Borgmann describes the difference between natural and cultural information at one point as follows: "Natural information emerges of itself, intimates rather than conveys its message and disappears. Cultural information, to the contrary, is wrested and abstracted from reality, carries a definite content, and assumes an enduring shape" (59). A price is paid for cultural information: "When literacy invades the oral culture, it drains vitality from the community" (51). Natural information, of course, never becomes obsolete, and remains important—but cultural information takes its station alongside, setting itself up as information for reality. Cultural information, as it were, is a recipe for reality, akin to an architectural blueprint or a musical score. In this account of cultural information, Borgmann describes reading fondly: "Thus to read is to gather our past and illuminate our present. It is a focal activity that collects our world and radiates back into our world. Reading at its best realizes a world view. It is solitary and outwardly passive. But in reality it vigorously engages and shapes our vision of the world. Intelligent reading of fiction and poetry, far from being an escape, is a tacit conversation with actual reality" (92). Borgmann does not disguise his deep appreciation for cultural information: "[N]othing so concentrates human creativity and discipline as the austerity of cultural information, provided the latter again is of the highest order, consisting of the great literature of fiction, poetry and music" (220).

Still, Borgmann esteems oral culture highly. Oral culture has its own value, which is especially evident when compared with literary culture. Borgmann speaks of cultural information in the beginning of modernity in quite ambivalent terms: "Things have the richness and particularity of a picture while conventional signs have the precision and generality of a concept. When meaning began to decline early in the modern era, the profuse ambiguity of natural things and works of art came to compare poorly with the austere definition of printed information" (114). Borgmann claims in fact, following Victor Hugo, that "the latter in fact hastened the demise of the former" (114). Again and again he repeats his view that reality has become silent, though poets can still be its advocates. Before, reality could speak to everyone; today, it speaks only to a select few.

Borgmann makes it clear that the intrusion of cultural information into the world of natural information has changed the human world. This is even more true in the case of *technological information*, Borgmann's third type of information. Technological information comes to be independent of reality—a tendency that already shows up in the case of cultural information—and ultimately appears as something that rivals and even replaces reality (182). This information is not *about* reality (natural information), is not a *recipe for* reality (cultural information), but it appears *as* reality (technological information). Musical performances sold in the form of CDs are a striking example of this—of technological information passing as reality. The thought that technology represents a specific way of dealing with reality is a classic Borgmannian theme. In information technology this is carried to an extreme. Such technology not only overcomes the resistance of reality, which we encounter in our everyday dealings but above all in human labor; it makes short shrift of it—liquidates it. In the word "hyperreality," used as a synonym for virtual reality, Borgmann detects the sense that virtual reality is superior to reality itself. This wedge between genuine (actual) reality and virtual reality has brought about a new ambiguity in contemporary culture. Genuine reality threatens to become only a part of cyberspace. "Supernatural brilliance, limitless variety, and unreal availability constitute the normative identity and charm of virtual reality. The actual world seems drab, poor, and hard in comparison" (187). This is meant ironically, of course; Borgmann is no believer in the "glamour and trivialization" that virtual reality offers us. He even holds it to be dangerous in that technological information blurs the boundary between fact and fiction, and "compared with the vividness and interactivity of actual reality virtual reality turns out to be a pale and brittle world and is bound to remain so" (198). However much one tweaks it, technological information leads, culturally, a parasitic life: "It draws much of its life blood from real and traditional culture" (199). Borgmann is as at war with "the noisy rhetoric of so-called radical innovation and liberation" and shudders at the thought that "this new age will change forever the way an entire nation works, plays, travels, and even thinks" (203). His remedy is to "hold on to reality," because it alone "remains at bottom inescapable and unfathomable. It is the ground on which ambiguities of the technological information can be resolved and its frailties repaired" (216). Borgmann's greatest concern is the damage that culture suffers thanks to information technology (221). The only possible way to address the resulting cultural crisis, he claims, is to right the balance of information and reality. "It amounts to the restoration of eminent natural information. A well ordered realm of natural information in turn is both hospitable to practices of realizing cultural information and enlivened by such practices. As for technological information, it is best allowed to develop freely within a world whose natural and cultural ecologies are guarded and engaged in their own right" (221).

A PERSONAL LOOK BACK

Borgmann's interpretation of technology and technological culture is rich and inspired, and offers a complete research program for the philosophy of technology that can be fruitfully explored. Borgmann has pursued it in connection with American society. The idea of the device paradigm is an extremely fertile one, and the different variations it can take make it even more interesting. In my own research on scarcity I have connected Borgmann's device paradigm with a theory of scarcity. This theory functions as a heuristic paradigm for interpreting modernity. In the realm of scarcity one wrestles with the inevitable shortage that arises due to the limitless desires of modern human beings. Technology claims to be able to solve this problem—indeed, to provide the only possible solution for it—insofar as its calling is the overcoming of this scarcity through the generation of endless amounts of infinitely varying commodities. But this solution is never finalized. New technological products fail to silence the demand for production, and instead stimulate new demands, with production and consumption leapfrogging each other. The two paradigms of scarcity and device thus go hand in glove. These insights may involve a slight transformation of Borgmann's theory; while he views the character of contemporary existence as stamped by technology, it may well be that contemporary existence is characterized more fundamentally by scarcity. In such a view, technology would appear on a scene *already* characterized by scarcity. Thus the scarcity theory might well provoke interesting clarifications of the dynamics of technological development.[3]

Borgmann's attempt to give an account of the implications of technological development for spirituality and for the orientation of everyday life is engaging. He is unequaled among philosophers in his ability to take a difficult notion such as focality and lay out its meaning in a very concrete and everyday manner. But while it is easy to subscribe to many of his critical observations about the technological revolution, his theory of focal practices on the micro- and macro-level, as well as his pleas for deictical discourse, are not completely convincing. He is surely right that technological culture is hardly hospitable to Christian faith, and even that it threatens to shut out all of its inspiration. The Christian concern for the powerless and the lost—the warm undercurrent of Christianity, Ernst Bloch calls it—has never been received with thunderous applause. But Borgmann's desire to use focal practices to make room for this tradition is surprising; trying to use focal practices to recover a disappearing engagement with reality is one thing, but it is quite another to be concerned with the care of "strangers, widows, and orphans." When Borgmann travels to upper Manhattan, he is encountering a community that cares for others. It is difficult to see how one can legitimately put this community in the same category as the Missoula baseball players, even in the terms of Borgmann's own descriptions. The celebrations that take place in the stadium in Missoula are completely different from those in the cathedral of St. John the Divine in upper Manhattan. The celebrations move in opposite directions: the first is oriented toward its own activity and glorification, while the second is oriented away from its own activity to the sphere in which "the left hand does not know what the right does," in the sense of that passage from the Sermon on the Mount where Jesus says, "But when you give alms, do not let your left hand know what your right hand is doing."[4] The winning team, that is, bids for the public's accolades, while Christian love does not call attention to itself.

In fact, it is precisely where Borgmann gives focal practices and celebrations a religious garnishing that he is least convincing to me. His descriptions of communal experiences couched in sacred terms raise numerous questions. For instance: which divinity is being celebrated? An exposure of these manifestations and mass experiences seems more adequate.[5] This is no small theological point, but has potential repercussions for the philosophy of technology as expounded in Borgmann's latest book.

The three types of information—natural, cultural, and technological—are structured differently. The ultimate criterion for judging the worth of information appears to be "reality." Natural information comes off well in this respect, cultural information a little less well thanks to its abstractions, while technological information receives the most critical treatment. The last two fare badly because the first is used as the measure. The *cultural* value of the natural information that he ascribes to hunters and gatherers he withholds from our technological culture. But each type of information needs to have its own standards of evaluation; they are distinct from each other, and cannot be judged in the same way. This does not mean, however, that I would deny the parasitic life of cultural and technological information.

Borgmann, I think, also speaks far too uncritically about natural information, and is far too accepting of religious declarations about reality. He tries not to restrict himself to a particular interpretation of the divine, and in his deictic discourse he chooses examples borrowed from different religions. This tolerance is very engaging, but as a good Bible reader Borgmann must be aware of how the children of Israel struggled with their beliefs and with those of their surroundings. Belief in Jahwe meant a constant struggle with the temptation of idolatry. The exclusivity, and the specific historical character, of the belief in Jahwe is revealed in the very first passage of the Ten Commandments.[6] The Jews had important historical events, such as the exodus out of Egypt, as constitutive for their belief. In other words, the one interpretation excludes the other. The ground state of information and reality is not univocal and unambiguously given. It is always disputed and contested. What holds for natural information also holds for the other types.

Each type of information, I said, requires its own criterion; moreover, cultural and technological information need to be treated as new creations in their own right. In his discussion of cultural information Borgmann implicitly relies on a standard of evaluation other than "holding on to reality"—and his evaluation of technological information turns out to be an implicit moral judgment. That not all that glitters is gold is self-evident in the technological world. Philosophers need to carry out an independent and adequate evaluation of the three types of information. And while I think it is clear that Borgmann's theory provides a superb beginning, more work needs to be done.

NOTES

1. Albert Borgmann, *Übereinkunft und Auseinandersetzung. Die ontologische Implikationen und Gundbegriffe in fünf von Newmans Schriften* (Freiburg im Breisgau, 1964).

2. In Holland, former arch-enemies such as liberals and social-democrats have participated in the government for several years now, and maintaining the governmental machinery has become their principal concern. The 1999 economic reform of Chancellor Schroeder in Germany is an austerity policy. An election promise was made to create new jobs, but that July the government imposed a steep cutback in expenditures. Its so-called reform amounted to a tax reduction for corporations and businesses, and a cutback in pensions was billed as a modernization of the social system. Schroeder's program is plain neoliberalism and has little to do with social-democracy. The Socialist Party of Lionel Jospin in France is a little different in that Jospin does not consider neo-Keynesian techniques outdated. But Great Britain's Tony Blair represents the neoliberal

"third way," which holds that markets are more intelligent than governments and that entrepreneurs are better informed than politicians; one hears no talk from him about the inability of markets to optimize economic and social benefit.

3. See my articles, "Scarcity, Labor and Modern Technology," in *Research in Philosophy and Technology: Technology, Ethics, and Culture*, vol. 17, ed. Carl Mitcham (Greenwich, Conn.: Jai Press, 1998), 81–106; "Desire, Technology and Politics," *Contagion* 6: 85–95; "Philosophy in the Service of People," *Technology in Society* 21, no. 2 (1999): 175–89. But Borgmann has remained skeptical of scarcity theory; see "A Scarcity of Focal Things: Reply to Pieter Tijmes," *Technology in Society* 21, no. 2 (1999): 191–99.

4. See, for instance, the poem "Christen und Heiden," in *Widerstand und Ergebung* (Munich: Chr. Kaiser Verlag, 1961), 246.

5. See *Sport-Eros-Tod*, ed. Gerd Hortleder and Gunther Gebauer (Frankfurt am Main: Edition Suhrkamp, 1986), 1335.

6. "I am the Lord thy God who brought you out of the land of Egypt, out of the house of slavery; you shall have no other gods before me." Relevant in this context is Herbert Butterfield's book *Christianity and History* (Glasgow: Fontana Books, 1960).

BIBLIOGRAPHY

Borgmann, Albert. 1974. *The Philosophy of Language: Historical Foundations and Contemporary Issues.* The Hague: Nijhoff.

———. 1984. *Technology and the Character of Contemporary Life: A Philosophical Inquiry.* Chicago: University of Chicago Press.

———. 1992. *Crossing the Postmodern Divide.* Chicago: University of Chicago Press.

———. 1993. "Finding Philosophy." In *Falling in Love and Wisdom,* edited by David D. Karnos and Robert G. Shoemaker, 157–60. New York: Oxford University Press.

———. 1996. "Technology and the Crisis of Contemporary Culture." *American Catholic Philosophical Quarterly* 70: 33–44.

———. 1999. *Holding on to Reality: The Nature of Information at the Turn of the Millennium.* Chicago: University of Chicago Press.

2. Hubert Dreyfus: Humans versus Computers

Philip Brey

HUBERT DREYFUS, CRITIC OF ARTIFICIAL INTELLIGENCE

In 1956, a mere ten years after the invention of the first programmable digital computer, the birth of a new field of research called *artificial intelligence* was announced at a conference at Dartmouth College in New Hampshire. Artificial intelligence, or AI (as the field soon came to be known), was described as a basic science that would systematically study the phenomenon of "intelligence." Its researchers would pursue this goal by using computers to simulate intelligent processes, and its point of departure was the assumption that the logical operations executed by computers could be structured to imitate human thought processes. AI researchers supposed that it was possible, in principle, for computers to be supplied (through proper programming) with genuine intelligence the same way that artificially produced diamonds are nevertheless genuine diamonds. Because the workings of a computer are understood while those of the human mind are not, AI researchers hoped in this way to reach a scientific understanding of the phenomenon of "intelligence."

From the very beginning, AI was a field with high goals and lofty promises. The highest goal was no more and no less than to construct a computer system with the intelligence and reasoning ability of an adult human being. Many AI researchers claimed that this goal would be reached within only a few decades, thanks to the invention of the digital computer and to key breakthroughs in the fields of information theory and formal logic. In 1965, the noted AI researcher Herbert Simon predicted that, by 1985, computers would be able to execute any task that human beings could. The equally famous AI researcher Marvin Minsky predicted in 1967 that all of AI's important goals could be realized within a generation.

It is easy to understand why such predictions were taken seriously at the time, given the apparently limitless possibilities that the computer seemed to offer. In addition, a string of early successes by AI researchers helped to legitimize the ambitious claims. In 1956, its first formal year of existence, AI scored its first victory, a computer program able to play chess at a novice level—and chess-playing programs improved steadily in ability almost every year thereafter. Other breakthroughs followed shortly. In 1964, a program called STUDENT was able to interpret, understand, and solve short textual passages containing algebra problems; two years later, ELIZA was able to carry out a modest therapeutic dialogue with people about their personal problems. Funding agencies took note, including the U.S. Department of Defense. Bright young researchers flocked to the new science. This inaugurated a huge growth spurt for AI, during which it established itself as an exciting, well-funded field in which hundreds of millions of dollars were spent annually worldwide, spread out among thousands of AI researchers.

As the 1960s opened, when AI was still a new field, a young philosopher named Hubert Dreyfus was introduced to it in an indirect way. Dreyfus, at the time an assistant professor at the Massachusetts Institute of Technology (MIT), was teaching courses on philosophical theories of knowledge and perception—but his students were telling him that the theories he was teaching had been rendered obsolete by the invention of the computer. Under the leadership of his colleague Marvin Minsky in the MIT Artificial Intelligence Laboratory, the students informed Dreyfus, MIT researchers were on the way to creating a machine that would be able to know and perceive on its own.

Goaded by this news, Dreyfus began discussing computers and their operations with his brother Stuart, who was then working as a computer specialist for the RAND Corporation, a prominent nonprofit research organization. Via this contact, RAND recruited him as a philosophical consultant to evaluate their new AI program. This program was headed by Allen Newell and Herbert Simon, who would later become famous for their AI work. But in his evaluation of their research, Dreyfus came to the conclusion that, while it had successfully demonstrated the ability of computers to solve certain specific types of problems, it had not provided any evidence of insight into the phenomenon of intelligence itself, and indeed was on completely the wrong track in seeking to simulate human intelligence. His pessimistic report, written in 1964 and entitled "Alchemy and Artificial Intelligence," was vigorously criticized by Newell and Simon, but was released by the RAND Corporation the next year over their objections.

Dreyfus's report was the first detailed critique of AI to be published, and almost immediately occupied the center stage of a heated debate among computer scientists worldwide. It was his first influential publication on the subject, and was the first of a series of philosophical critiques of AI in the form of books and articles. Dreyfus's most important publication in this area is *What Computers Can't Do,* which brought him international fame as a critic of AI. First published in 1972, it was substantially revised and republished in 1992 as *What Computers Still Can't Do.* Another important contribution to the AI discussion is *Mind Over Machine* (1986), co-authored with his brother Stuart.

A remarkable aspect of Dreyfus's critiques is that they are motivated by a philosophical tradition—phenomenology—which at the time was not often associated with science and technology and seemingly far removed in its concerns. Phenomenology, as it appears in the work of Martin Heidegger and Maurice Merleau-Ponty, applies itself to describing the interrelationships between human beings and the world, and uses the first-person experiences of human beings as a point of departure. And while Heidegger, Merleau-Ponty, and other phenomenologists have quite specific things to say about the nature of human perception, thinking, and behavior, their pronouncements about science and technology tend to be rather general and abstract. Dreyfus, however, was able to apply their ideas skillfully in his critique of AI to reach quite specific and concrete conclusions.

Ever since his earliest work on the subject, Dreyfus has progressively honed and extended his philosophical critique of AI by broadening his use of the work of phenomenologists such as Heidegger, Merleau-Ponty, and Husserl, and by making use of the insights of other philosophers, including Michel Foucault and Søren Kierkegaard. One of Dreyfus's principal concerns, which appears with regularity throughout his writings, is to articulate the various ways in which human beings experience the world and develop manners of getting around in it. One important book in which he takes up this theme is *Being-in-the-World* (1991), considered one of the best and most lucid expositions of Heidegger's early work.

Another regularly recurring concern is his critique of Cartesian rationalism. In Dreyfus's interpretation, the key rationalist assumptions are that reality has a rational structure built up out of independent elements in a rule-governed way, that human thinking works in the same rational manner, and that everything that is not rationalizable—that cannot be expressed and defended in rational principles—is of little if any value. Dreyfus is convinced that Western culture is still shaped to a large extent by these rationalistic assumptions, but he is also convinced, based on his readings of Heidegger, Merleau-Ponty, and Wittgenstein, that these assumptions are fundamentally flawed. Rational, formal structures are, according to him, human constructions that are only subsequently imposed on reality. Knowable reality itself lacks a rational structure; its features are co-determined by human needs and actions. The most fundamental way of knowing is intuitive rather than rational. Rationalism, as it crops up in AI and elsewhere, knows nothing of these original structures of reality and fails to do justice to the role of intuitive knowledge and skills. Dreyfus is an unrelenting advocate of intuitive knowledge and skills, and a sharp critic of rationalism in all of its contemporary guises.

Dreyfus's critique of AI has been influential to a degree that is exceptional for a philosopher. He has acquired a reputation among AI researchers—though initially, at least, a negative one—as well as among philosophers interested in AI. His works are studied by many non-philosophers in computer science and other technical fields. But most remarkable of all, many influential AI researchers have taken Dreyfus's critiques to heart, and developed and applied many of his phenomenological ideas to their own research. Via this route, the frequently abstract philosophical notions to which Dreyfus appeals have had a direct impact on the development of AI.

This presentation of Dreyfus's work will begin with an outline of classical AI research. This will be followed by an exposition of Dreyfus's critique of classical AI; of his alternative theory of intelligence; of his critique of neural networks, an important recent approach to AI; and of his critique of the societal applications of intelligent computer systems. Finally, I shall evaluate the soundness and influence of Dreyfus's critique.

The Classical Paradigm of Artificial Intelligence

From the beginning there have been many different types of AI research with different goals, methods, and formalisms. Yet also from the beginning of AI research in the 1950s up to the beginning of the 1980s the different types of AI research had so much in common as to constitute a paradigm, in the sense articulated by philosopher of science Thomas Kuhn: a collection of methods, goals, assumptions, and exemplary examples of successful research that are shared by scientists and that together define a research program. This paradigm, which continues to characterize much AI research, is known by different names, but I shall refer to it as "symbolic AI" (or sometimes as "classical AI") because its central tenet is that intelligence is symbol manipulation.[1]

In the first decade of its existence, symbolic AI had as its goal the construction of intelligent computer systems. The grail was a system that possessed universal intelligence—that is, had a universal ability to reason, solve problems, understand language, and carry out other intelligent tasks the way an intelligent human adult could. This research was initially not directed at developing technical applications, and was promoted above all as a science: the new science of intelligence. Some classical AI researchers, including Newell and Simon, set as the explicit goal of their research the modeling of the cognitive (thought) processes of human beings; AI with this goal is sometimes called "cognitive simulation." Within this approach, AI programs are considered to both simulate and explain intelligent human behavior.

Other researchers who do symbolic AI, including Minsky, do not pretend that their computer programs simulate human thought processes, but rather that their work provides a theoretical contribution to the understanding of the phenomenon of "intelligence" by laying bare the general properties of intelligent processes. They claim that while their research may not allow direct insight into the psychological performance of intelligent tasks, it does allow insights into the performance of competent intelligent human behavior; that is, it provides general insights into the cognitive abilities that human beings must possess in order to display intelligent behavior. But the differences between the approach of cognitive simulation and this more usual approach are of less importance than their points of agreement; both approaches take as their goal the understanding of the phenomenon of intelligence, and they share important theoretical points of departure, methods, and formalisms.

Symbolic AI assumes, as its principal point of departure, that intelligence is a matter of manipulating symbols following fixed and formal rules. A series of assumptions is made to arrive at this idea. A first, necessary assumption is that all intelligent processes, including perceiving, reasoning, calculating, and language use, are forms of *information processing*, that is, of uptaking information from the environment, processing or manipulating this information, and providing a response. Thus when one adds numbers, one first determines the addend, performs a certain operation on this information, and then exhibits the solution. Chess playing, somewhat more sophisticated, nevertheless has the same structure; one inventories the pieces and their positions, analyzes the situation, and determines which move is to be made. The implication of such examples seemingly is that intelligent organisms and systems have in common that they are *information processing systems.*

Once this assumption is made, one is naturally led to pose the following two questions: what is the character of this information, and how is it "processed"? At this juncture symbolic AI makes two key assumptions. In response to the first question, it assumes that, to be processed by an information processing system, information must first be *represented.* In order to handle information, a system must first operate with a medium in which the information can be entered. Such a medium, which provides information about an external reality, is called a *representation.* Familiar examples of representations include photographs, pictures, images, and spoken and written sentences—but these are not the kinds of representations that can be used by information processing systems. Information processing systems are assumed to use *internal* representations, all given in a form adapted to what they can handle. Thus human thinking is supposed to work via a system of internal mental representations in which our thoughts, perceptions, and memories are all inscribed.

The most fundamental assumption of classical AI is that the internal representations of intelligent systems are symbolic in nature. An alternative possibility is that internal representations are more like photographs and images in that they are iconic, carrying information by a physical resemblance with that to which they refer, the way a portrait conveys information about its subject by virtue of its likeness. But classical AI takes as its starting point the assumption that the internal representations are more like the words of a natural language. Language is symbolic; its tokens are arbitrary in the sense that they have neither resemblances nor inherent references to that which they represent. The word "dog" neither looks like a dog nor has any intrinsic connection to one.

The assumption that whatever bears the information in intelligent systems is symbolic is more convincing than the assumption that it is iconic. It is difficult, for instance, to imagine an iconic representation of abstract things with no perceivable structure. Symbolic representations are also much easier than iconic ones to combine, and a finite number of symbols can be used in different combinations to represent an infinite amount of content. Natural language is a case in point; a finite number of words can be combined to create, in principle, an infinite number of sentences. Also, symbols clearly play a prominent role in the sorts of cognitive tasks, like mathematical calculations and logical reasoning, that are often taken to represent the highest form of intelligence.

From the earlier assumption that intelligence consists of the capacity to process information, together with the assumption that information processing consists of symbol manipulation, it follows that intelligent systems are *symbol processing systems*. Up to now nothing has been said about how these symbols are processed. It is a further assumption of symbolic AI that symbols are processed only on the basis of their *formal* properties, that is, the form of the symbol as opposed to its content or meaning. The meaning of a symbol, therefore, does not play a direct role in its processing. Thus when a computer processes the symbol "dog" in a particular way, it does so because of what it recognizes in the form of the symbol and not because it has any insight into the meaning of the symbol.

But how is it determined which processes are carried out by the system on the basis of these formal properties? Here we encounter another key assumption: an information processing system includes *rules* according to which these symbols are interpreted and processed. These are of necessity formal rules, because they are linked only to the formal properties of these symbols. They work automatically, as it were; when presented with a symbol or series of symbols, the system executes a certain process that results in a new symbol or set of symbols, which then once again is linked automatically to another rule, and so forth. In the absence of such rules, intelligence would be a mystery, at least from the perspective of symbolic AI, because there would be no easy way of accounting for the symbol-processing ability of intelligent systems.

More support for the assumption that intelligent information processing consists of the application of rules comes from the role that rules appear to play in many tasks requiring intelligence. Thus language comprehension appears to involve knowledge of rules of grammar, logical reasoning appears to involve the application of analytical rules, and problem solving in mathematics and natural science appears to involve the application of mathematical principles or natural laws. Notice the assumption that knowledge required for intelligent behavior is theoretical; to know and understand something means to possess an abstract, symbolic theory expressed in rules with which one apprehends the phenomenon.

The theory of intelligence just sketched can be formulated and elaborated without reference to the nature and possibility of the digital computer. Clearly, however, the development of the computer makes this theory of intelligence considerably more attractive. A digital computer is conceived to be an information processing system that makes use of symbolic representations (strings of zeros and ones) and that processes these symbols according to formal rules (a likewise symbolically represented *program*). The computer's existence therefore seems to offer the opportunity to test and explore assumptions about the symbolic and rule-governed character of intelligence and build models of intelligent processes in a scientific manner. The above-mentioned assumptions about intelligence thus offer the prospect of a potentially fruitful scientific research program that might well lead to interesting technical applications. The initial successes of symbolic AI in the design of intelligent computer programs seem to supply an additional legitimacy.

The ideas about the nature of intelligence promoted by symbolic AI are sometimes described as innovative, but Dreyfus emphasizes that they are merely the latest reincarnation of an ancient view, generally called *rationalism*, which periodically emerges in the history of thought. Plato laid the groundwork for this theory. Plato proposed that wisdom consisted of the ability to formulate knowledge in explicit definitions, and scorned human beings whose behavior was based on learned abilities or intuition. He believed in the possibility of discovering a system of theoretical, objective principles that, much like the fundamental axioms of geometry, could be used to justify behavior and explain reality on a rational basis.

The most important representative of this idea in the modern era is René Descartes, who in the seventeenth century claimed that each problem can be broken down into simple and independent elements, and that every complex situation or thought can be clarified by discovering the system of rules that govern how this situation or thought has been built up out of these simple elements. He claimed that even the human mind operates according to such rules and simple elements. After Descartes, this conception of the nature of intelligence crops up again in other rationalistically minded thinkers, including Leibniz, Kant, and Husserl, but also to a lesser extent in empiricists such as Locke and Hume, and more recently in scientists such as linguist Noam Chomsky, the psychologist Jerry Fodor, and the different representatives of symbolic AI.

According to Dreyfus, three characteristic notions of rationalism can be found in symbolic AI. The first he calls the *psychological assumption,* the already-mentioned supposition that human intelligence is a question of symbol manipulation according to formal rules. This assumption provides the theoretical basis for the claim that a computer can be programmed to think like a human being. Not all AI research programs share the psychological assumption, though many do; it is the explicit point of departure, for instance, of the cognitive simulation approach. All variations of symbolic AI, however, share a second, *epistemological assumption,* that all knowledge is formalizable; that everything that can be understood by human beings can be expressed in context-independent, formal rules or definitions. If true, this supposition would guarantee the success of the project of symbolic AI even if the psychological assumption is false, because a formalized version of informal human knowledge and behavior would have the same cognitive value as the unformalized original. While a computer might not be in a position to simulate human thought processes, it would be able to reproduce human intelligent behavior.

Both the epistemological and the psychological assumptions are frequently based on the *ontological assumption* that reality, insofar as it can be known by human beings, has a formalizable structure that is built up out of a series of objective, determinate elements, each of which is independent of the others. If reality lacked such a structure, it is highly unlikely that it would be knowable with the aid of a set of context-independent, formal rules whose existence is taken for granted in both the epistemological and the psychological assumptions.

The fact that classical AI is based on a rationalistic conception of intelligence that belongs to the history of philosophy would not be of interest except for the fact that that history also includes important philosophical critiques of that conception. Dreyfus has been heavily influenced by these anti-rationalist critiques, especially those of Heidegger, Merleau-Ponty, and Wittgenstein, and he makes extensive use of their arguments in his critique of symbolic AI and its three central assumptions, as well as in his development of an alternative theory of intelligence.

DREYFUS'S CRITIQUE OF SYMBOLIC AI

The two scientific fields that could furnish proof in support of the psychological assumption of classical AI are psychology and neuroscience. Neuroscience is important because thinking with the aid of rules and symbols would be possible only if these rules and symbols were implemented in the human brain in the same way that a computer program is implemented by the hardware of a computer. Dreyfus calls the assumption that brain functioning involves the implementation of a symbol processing system the *biological assumption,* a fourth assumption that frequently appears in classical AI. But Dreyfus emphasizes that both psychological and brain research have not so far provided good empirical evidence for the psychological and biological assumptions, though neither is there any conclusive proof for the incorrectness of these assumptions.

Dreyfus's most important criticism, however, is directed against the epistemological assumption, underlying all forms of classical AI, that intelligent behavior can be reproduced by formalizing human knowledge (i.e., codifying it in rules) in a way that can be followed by a machine. Dreyfus argues against this assumption, stating that while such formal rules may be one way of *describing* human knowledge, they cannot provide the basis for *reproducing* such knowledge. The possession of knowledge, Dreyfus points out, entails the ability to apply this knowledge in relevant situations by applying it in reasoning, communication, and behavior. The idea that fire is hot, for instance, entails being able to apply this knowledge at appropriate moments in thinking about or dealing with fire; if this didn't take place, one could not truly say that this knowledge was present.

The application of formalized, rule-given knowledge, however, appears to run up against an important problem. If a computer that has been given a set of formal rules is to be able to apply them to a new datum—for example, to a typed-in sentence, an arithmetic sum, or a photographic image—it must first be told under what precise circumstances they are indeed applicable. This may seem to be simply an issue of symbol matching: if the datum (encoded in symbolical form) has the appropriate formal properties referred to in a rule in the computer program, then apply the rule. Almost invariably, however, it turns out that contextual factors also play a role in rule application. Consider, for instance, the attempt to program a computer to understand language. The simplest approach would be to give the computer a set of rules of interpretation that define the meanings of different words and a set of rules of grammar to analyze the sense of the sentences. The computer would then recover the meaning of strings of text by applying these two sets of rules.

But numerous problems arise here, one being that many words are ambiguous. Suppose for instance that the computer is given the word "hot" in a sentence and is asked to describe its meaning. One rule that might apply in this case has the form, "If something is hot, it has a high temperature"; but a second possible rule has the form, "If something is hot, it has a sharp peppery taste and will burn the mouth." In order to know which of these two rules of interpretation is to be applied, other elements are relevant, for instance whether the text refers to food. Higher-order rules must therefore be formulated for the correct application of the rules of interpretation; "If in the preceding text, reference is made to a peppery dish, then apply the second interpretive rule." But there are also exceptions to these rules of application. A text can be about hot Mexican dishes, but nevertheless use the word "hot" to refer to something with a high temperature. Thus there need to be still higher-order rules of application for the appropriate use of the rules of application—threatening to give rise to an infinite regress of rules, making interpretation impossible.

In short, the problem appears to be that the correct interpretation of many elements depends heavily on surrounding elements. Formal rules, however, need to be tied as little as possible to the context and to be related only to the elements in question, or to a few that are important in determining its application. If one sought to make rules sensitive to context, all possible contexts would have to be formulated, or separate rules of application would have to be formulated. Both solutions appear to be without an end.

Human beings, Dreyfus observes, *are* able to interpret elements effortlessly from the context. Thus if they encounter a misspelled word in a text, they automatically fill in the right meaning, while computers grind to a halt. Human beings, Dreyfus concludes, have "common sense," by virtue of which they know which interpretations are meaningful and which not. Computers lack common sense, which is why they often reach absurd interpretations. Dreyfus calls providing computers with common sense the greatest challenge of classical AI, and calls it *the common-sense knowledge problem* of classical AI. But it is a problem Dreyfus considers insoluble, for the reasons mentioned above.

Computers function the best when the "world" that they encounter and interpret is an artificial and formal world. A formal world consists of elements whose identity can be directly read off from their form, independently of whatever other elements exist in the world, and that are linked to these other elements in clearly ordered ways. In such a formalized reality the common-sense knowledge problem rarely crops up. This is the case, for instance, in simple games like tic-tac-toe and to an extent even in complex ones like chess, and also for mathematics and formal logic. The problems encountered in such knowledge domains are nearly always characterized by a clear goal or "final state" to be reached—three X's in a row, checkmate, or a the numerical value of a mathematical equation—as well as by a set of clearly defined steps to be taken in order to reach the goal. And in fact classical AI has had its greatest successes in solving just these kinds of problems.

But more mundane problem situations involving intelligent behavior appear to have an entirely different structure than that of these artificial "worlds." Let's take for instance the problem situation that occurs when you accidentally lock your car keys inside the car after a trip to the supermarket. Clearly this situation involves a problem that requires a solution, though the problem differs from those mentioned in the previous paragraph on two counts.

The first difference is that the goal related to this solution may not be clear in advance. Clearly, you are posed with a problem that you will have to overcome, but what is your goal? Is it to regain access to the car keys in the car? Not if you have access to a set of spare keys that a friend can readily bring over. Is it to be able to drive the car again? Not necessarily if your greatest need is to return home as rapidly as possible. The search for a solution is not directed toward a single, unique goal, but involves a continuous weighing of different needs, including the estimated damage to the automobile, the cost of lost time, returning home or making appointments promptly, and so on.

The second difference is that, even when the goal is clear, the problem is not readily formalizable because it is not apparent in advance which facts are potentially relevant to finding a solution. The situation, that is to say, is not characterized by a fixed set of elements with objective properties and characteristics to which rules can be applied. Some facts become potentially relevant only during the actual solving of the problem, such as the car window that turns out to be ajar or weakly fastened, or a wire hanger lying on the street that might be used to break in, or a previous owner who just might happen to have kept a set of reserve keys. The process of solving this kind of problem, that is, typically progresses through several stages in the course of which one conceptualizes and reconceptualizes it in search of a representation of the problem that gives one the best feeling that one has a confident grasp of the situation. The ability to creatively reformulate a problem appears to be a more essential skill than the ability to find a solution to a problem whose definition is clear and well defined from the outset. Formal rules and their application appear to play no role in the search for a good definition of the problem.

In short, there appear to be sound arguments against the epistemological assumption that intelligent behavior can be reproduced by a system consisting of formal rules and symbols. Moreover, human intelligence itself does not seem to work in this way.

INTELLIGENCE IS EMBODIED AND SITUATED

In his alternative theory of intelligence, Dreyfus argues that one must begin by recognizing that human beings generally do not apply rules in their intelligent behavior—and generally do not even make use of internal representations. For Dreyfus, intelligence is *situated;* it is co-determined by the situations in which human beings find themselves. The insights on which intelligent behavior is based are constructed locally, from concrete situations, with the aid of information that is a direct product only of this situation and without the aid or necessity of prior rules or internal representations. This vision, which derives mainly from the philosophy of Heidegger—and to a lesser extent from that of Wittgenstein and Merleau-Ponty—is probably the most difficult part of Dreyfus's work to understand.

The psychological assumption that human beings need representations and rules in order to interpret the world seems so reasonable, according to Dreyfus, because of a particular conception of how the world is and how it is known by human beings. The world, in this view, is interpreted as a material structure, independent of human beings. This is just the view offered to us by the natural sciences. This world is inherently meaningless, and is spatially separate from human beings, so that no direct, unmediated experience of it is possible. Dreyfus does not deny the value of the perspective offered by the natural sciences about the world, but emphasizes that another perspective is possible—the phenomenological perspective. In its description of the world, this perspective takes human experience as its point of departure. The 'world,' as the word is used by phenomenologists, thus refers to the world as it is manifested in human experience.

This "human" world is a world that is not entirely objective, as it is filled with experienced structures, like smells, feelings, frustrations, threats, obstacles, and goals. Nor is it completely subjective, in the sense that the structures that we learn to perceive in the world are not our own arbitrary mental constructions; smells and obstacles are not things that we invent but are manifested in our encounters with the world. Human beings are born into, interact with, and learn to perceive, behave, and think in a world that is neither entirely objective nor entirely subjective. This world evolves alongside these activities, for new worldly structures are always manifesting themselves in and through human activity. Thus while the world of a newborn baby is to a great extent unstructured, that of an adult human being contains countless structures that have crystallized out in the course of years.

Dreyfus emphasizes that, for human beings, the experience of the world as a whole precedes the experience of independently distinguished elements. Thus a depressed person experiences the world as "gray" and "meaningless" before specific elements stand out in it, and one may experience a new environment as "safe" or "threatening" before distinguishing discrete objects; it is the situation as a whole that calls for the experience. Specific elements in a world or situation are distinguished and experienced from out of this more general experience of meaning and sense. As a result, these elements stand always in a meaningful relation of significance with their context. While at work, a carpenter experiences a hammer that lies close at hand as a "thing-with-which-to-hammer" and as a "thing-that-is-useful-with-nails," but in a more threatening context might experience it as a "weapon-to-use-against-an-intruder." In neither case does the carpenter perceive the hammer as just one item in the environment among all the rest, whose meaning and significance are still to be determined. In the same manner, a chess grandmaster "sees" a meaningful board situation and its associated possibilities, without having to first build it up by inventorying the locations of specific pieces and surveying their possibilities as allowed by the rules.

Intelligent behavior involves human beings discovering meaningful structures in situations in which they find themselves, which call in turn for meaningful behavior. The meaningful structure that human beings find in such situations is a local product of their needs, actions, and perceptions. The different elements in this structure derive their meaning from it. Actions flow automatically from the meaningful context; just as the eye automatically "understands" the amount of light it receives and reacts by increasing or shrinking the size of the pupil, so human beings "understand" the situations in which they find themselves and react accordingly with actions appropriate to the context. These actions can be generated fairly automatically out of the experienced situation because these situations are already structured in "manageable" ways, that is, with an eye to meaningful behavior.

The meaningful structure that is experienced in a situation is thus not one that has been built up according to the application of a number of fixed rules out of separate, context-independent elements. If it were, the assumptions of symbolic AI would be sound. But the reality is exactly the opposite: the global, holistic structure that belongs to each situation makes it possible, by a process of abstraction, to discover and represent elements in it as separate objects and facts, and then to apply rules to them. For intelligent behavior it is usually not necessary to abstract in this way, except when the problem situation is defined in abstract ways from the outset.

Dreyfus's views about the situatedness of human behavior form one major part of his theory about human intelligence; the other major part consists of his view that intelligence is *embodied*—that is, requires a human body (Dreyfus 1967, 1972, 1996). This view, which is not entirely independent of the first, derives mainly from the philosophy of Merleau-Ponty. However, Dreyfus's explication of this view is, much like his explication of his ideas about the situatedness of intelligence, often unclear and schematic. It is unclear, for instance, whether Dreyfus means that intelligence is something that is distributed throughout the entire body and thus cannot be spoken of as localized in the brain or the mind, or whether he means that intelligence can exist without a body, but can be developed only with the aid of a body. Let's consider the plausibility of each in turn:

1. Does Intelligence Require a Body?

Scientists as well as nonscientists often assume that intelligence is localized in the brain. However, in the case of at least one important type of intelligence, *sensorimotor intelligence,* this assumption is clearly disputable. Sensorimotor intelligence is the skill that human beings use in perceiving, recognizing, moving, and manipulating objects, as well as in coordinating and integrating perception and movement. The development of sensorimotor intelligence clearly requires a body, but this of itself does not mean that sensorimotor intelligence is also localized in the body; it is in principle possible that sensorimotor intelligence is exclusively a product of the brain in response to stimuli provided by the senses and then carried out by the musculature.

An alternate and equally defensible hypothesis, however, is that sensorimotor intelligence is localized in a complex feedback system that comprises the nervous system, the senses, the glands, and the muscles. All these elements could then be analyzed as information processing systems, or parts thereof. Sensorimotor intelligence would then be a property of a fully developed body, in which not only brains but other organs pass through a training process leading to the development of a total system able to carry out intelligent and fully coordinated perceptions and movements.

But even if this hypothesis is correct, it is unlikely that all human intelligence is distributed in the body. Especially abstract, "higher" forms of intelligence, like abstract reasoning and calculation, do not appear to be dependent on a body. Human beings can have limbs and organs amputated or paralyzed and still not lose their ability to engage in abstract thought, and it is at least a theoretical possibility that, as sometimes depicted in science fiction stories, a brain could be removed from its body and kept in laboratory conditions while still retaining the ability to think. Not all types of intelligence thus appear to require a body. So if this is what Dreyfus means when he says that intelligence is embodied, his position is implausible when it is supposed to apply to "higher" forms of intelligence.

2. Can Intelligence Develop Only with the Aid of a Body?

Even if a body is not required for the *possession* of intelligence, it could still be required for the *development* of intelligence. It is obvious that a body is required for the development of sensorimotor intelligence, but for more abstract forms of intelligence this assumption is less plausible. An alternative hypothesis, compatible with the psychological assumption of symbolic AI, is that abstract intelligence is based on an innate symbol system in the brain, which in principle can develop independently of the body just as a computer does not require a body in order to extend its knowledge capabilities.

However, the first abstract thought processes that children develop appear to be closely integrated with their sensorimotor intelligence. Thus their first use of language is strongly tied to the world in which they perceive and behave, and their first use of numbers is related to concrete objects. Their imaginative abilities are also strongly associated with this sensorimotor world. Hence, an alternative hypothesis for the development of abstract intelligence is that it is not based on fundamentally new abilities but rather on abilities that are already involved in the development of sensorimotor intelligence.

Sensorimotor intelligence includes abilities such as pattern recognition, the mental grouping and manual manipulation of objects, the assessment of the impacts of forces on things, the visual taking apart and transformation of spatial structures, and the mental anticipation of the effects of actions. A developing abstract intelligence might be directly built up out of such abilities through their application to abstract domains. Thus even the manipulation of abstract symbols, as in mathematics and formal logic, ultimately would lead back to our ability to manipulate material objects in space and time. This is the view Dreyfus appears to lean toward, and he refers in his recent work to the studies of Mark Johnson (1987), who has tried to demonstrate that abstract concepts and abstract logic ultimately can be reduced to concrete, sensorimotor structures.

If intelligence is indeed situated and embodied, then it does not appear possible for digital computers to possess the broad scope of human intelligence, for they are not embodied and do not have a full human world at their disposal. The intelligence of computers appears to be limited to the performance of tasks in well-defined, formal domains and will fail in a complex human world.

THE NEW PARADIGM OF NEURAL NETWORKS

Not only have the shortcomings of symbolic AI become ever more apparent in recent years, but a rival AI paradigm has also arisen, called *neural networks* or *connectionism*. Neural network AI, which began to be developed in the early 1980s, is viewed by most researchers as a radical alternative to symbolic AI, rejecting from the start the idea that intelligent behavior springs from the manipulation of symbols according to formal rules. The neural network approach derives its inspiration for the modeling of intelligent processes not from the digital computer, but from the structure and operation of the human brain. What this approach still has in common with symbolic AI is that intelligence is regarded as consisting of information processing.

The structure and operations of neural networks are built to resemble those of the human nervous system, specifically the brain. The nervous system is built up out of nerve cells (neurons). Neurons can be conceived as tiny information processing systems: they receive stimuli from other nerve cells or sometimes directly from sense organs, and react by delivering electrochemical stimuli to other nerve cells, or sometimes to muscles and glands, where they are taken up by receptors. Whether a neuron delivers such stimuli and how strongly depends on a physiologically determined "program" in the neuron that responds to the way the impulses it receives reinforce or interfere with each other. When these impulses are above a certain threshold, the nerve cell reacts, and it delivers an impulse in turn to its surroundings. Neurons can therefore be conceived as processors with relatively simple input/output functions.

Researchers think that the difference between human nervous systems that differ in their intelligent capabilities is chiefly determined by the way the neurons in them are connected with each other and with the rest of the body. Intelligence is therefore mainly a product of the connections that the neurons enter into—hence the name *connectionism*. Neurons develop through entering into or breaking off, or strengthening or weakening, connections with their surroundings, depending on the way they are stimulated. At birth the connections that the nerve cells enter into are to some extent arbitrary, but as the infant interacts with the surroundings, its nerve cells adapt in such a way that the behavior they instigate becomes progressively more intelligent and successful. Therefore, to say that a nervous system learns means that the connections between nerve cells become modified by experience.

Neural network AI tries to create artificial intelligence by attempting to simulate the operation of the nervous system, through constructing a system of simple information processors with input/output functions resembling those of nerve cells. The number of processors can range from a few dozen into the thousands, and the strength or "weight" of their connections changes depending on the stimuli they receive. They typically consist of an input layer through which information is entered, one or more intermediate layers, and an output layer. In practice, thus far at least, neural networks are not true physical constructions but are simulated in ordinary digital computers. In recent years, however, these computers have become extraordinarily powerful, consisting sometimes of tens of thousands of parallel processing computers, and research is under way on parallel computers based on optical fibers.

Existing neural networks turn out to be astoundingly good at carrying out certain intelligent tasks, such as pattern recognition, categorization, and the coordination of behavior. Neural networks, for instance, have been built that are able to recognize human faces from different angles, and that can vocalize words on the basis of a written text. Neural networks give their best performance with tasks that require "lower" forms of intelligence, such as pattern recognition and categorization of perceptual stimuli. However, thus far neural networks have been unable to tackle tasks requiring the application of higher intelligence, such as mathematical or logical problems—which are precisely the problems with which symbolic AI has scored its best successes.[2]

Dreyfus has asserted that the basic assumptions of neural network AI are compatible with his own vision of intelligence (Dreyfus and Dreyfus 1988; Dreyfus 1992). Neural networks relinquish the rationalistic idea that intelligence is a matter of symbol manipulation and rule application. Knowledge in neural networks is not a matter of possessing explicit representations, but rather of the appropriate connections (ultimately) between nerves and muscles. Knowledge involves possession of an ability: it is more knowing *how* to do something than knowing *that* an assertion is true. In neural network AI, intelligent processes are frequently holistic and intuitive. Moreover, neural network AI is fully compatible with the assumption that intelligence requires a body and is situated: higher processes are often built up out of lower ones, and intelligence is conceived as something that develops through interaction with the environment. Thus by Dreyfus's own criteria, neural network AI appears to have more of what it takes to manufacture artificial intelligence.

But Dreyfus is ultimately pessimistic about the possibility that neural networks will ever realize this lofty aim. The problem lies not in the basic assumptions of neural networks, but in the incredible complexity of human intelligence. The fundamental problem with neural network AI is that, once again, the problem of "common sense" crops up, though in a somewhat different form. The intelligence of neural networks to a large extent depends on experience, for it is based on the connections that have been cultivated in order for the network to deal effectively with those situations that it happened to encounter in the past. The ability to deal intelligently with new situations depends on the ability to *generalize* intelligently from these past experiences to new ones. But here it seems as though intelligent criteria are first required in order to make intelligent generalizations: which past knowledge is relevant to the new situation, and which adaptations to it are needed in order to apply the knowledge to the new situation? If in the past I have eaten apples only inside, does this allow me to conclude that apples are also edible outside?

In principle, all the knowledge that a person possesses can be relevant in generalizing to new situations, and one cannot determine in advance what is and is not relevant. One must therefore have at one's disposal all the knowledge stored in one's brain in order to be able to generalize successfully. The same would be true of the generalizing ability of neural networks. But this suggests that a network that is able to generalize as successfully as a human brain would have to consist not of dozens or hundreds of processors, but of millions.

Aside from the fact that such networks are currently impractical, there is still the question of how such a network could possibly acquire all the relevant knowledge possessed by a normal human being. To acquire this knowledge would seem to require that the network pass through the same learning trajectory as an adult human being—but this would require that the neural network be embodied. It therefore appears that a neural network that could generalize as intelligently as a human being could happen only if it were built with the complexity of a human brain, supplied with an artificial body resembling a human one that passed through a developmental trajectory similar to that of human beings when they mature. But so far the creation of such an android form of life belongs to science fiction.

FROM MEDICAL SPECIALIST TO TEACHER: INTELLIGENT COMPUTER SYSTEMS IN SOCIETY

Although the promise of eventual practical applications may have played a role in the early enthusiasm of AI researchers and their supporters, few such applications had turned up by the end of the 1970s. But during the 1980s, things started to change and AI began to take on the character of a technology, as opposed to a theoretical science. The ultimate goal of most recent AI research is concerned with interesting technological applications. AI research has largely lost the ambition, which it had in the cognitive simulation approach, to make a fundamental contribution to the scientific explanation of intelligence.

Indeed, many AI researchers have stopped referring to themselves as scientists and now call themselves knowledge engineers.

AI technology has become a multi-billion dollar industry, and ever since the late 1970s has delivered a stream of interesting products such as chess computers and expert systems. And ever since the 1990s there has been an upsurge in conventional devices that have been equipped with artificial intelligence, such as "intelligent" vacuum cleaners, washing machines, and video cameras; control systems in industry; and "intelligent" computer software like the more sophisticated Internet search engines and operating systems that adapt their behavior to the user's habits. These kinds of applications mean that the boundary between AI research and other technological research—especially in computer science and electrical engineering—is rapidly disappearing.

The widespread use of intelligent computer systems in society has brought about new philosophical—and especially ethical—issues. Intelligent computer systems make *choices* and *decisions* according to criteria that the users generally have little or no understanding of. In effect, the computer systems take over responsibility for such choices and decisions from human beings. Handing over some decisions to computers—such as those involved in intelligent video cameras or in chess programs—is ethically unproblematic. But ethical issues are clearly raised when decisions are handed over to computer systems about issues like criminal punishment or whether to admit someone seeking political asylum.

The most important ethical problems related to intelligent computer systems are associated with expert systems. Expert systems, the first of which were developed in the mid-1970s, are computer systems that are intended to take over tasks from experts in a particular specialized domain; examples have been developed in medicine, law, industry, mathematics, science, financial planning, and accounting. Thus expert systems have been created to diagnose illnesses and recommend treatments, to track down flaws in airplane engines, to identify geological sites where valuable minerals might be mined, to put together investment portfolios, to establish whether a person deserves unemployment compensation, and to determine punishments for convicted lawbreakers.

Expert systems are mainly built according to the assumptions of symbolic AI. Their designers try to provide these systems with the required knowledge by interviewing experts and seeking to make explicit their often unverbalized and intuitive knowledge. This results in a list of often thousands or tens of thousands of facts and heuristics (rules that experts are thought to follow in reasoning) that are then translated into a computer program. The performance of the system is then compared with the performance of a human expert. If the system appears to perform satisfactorily, it can be put to use.

Despite his criticism of symbolic AI, Dreyfus was relatively optimistic in his early work about the prospects of expert systems. He had always claimed that computers could perform well in formalized domains that required little common sense. The sort of knowledge that experts like chess grandmasters and scientists acquire appears to be formalizable in rules, and appears to call for little common sense or everyday knowledge. Dreyfus claimed that computers in these specialized domains of knowledge might well be able to log striking successes.

But Dreyfus later reconsidered. What catalyzed this change of heart was a study he conducted together with his brother Stuart of the manner in which human expertise develops in a particular area (Dreyfus and Dreyfus 1986). This study seemed to show that humans employ rules in early stages of learning, but in later stages replace this with an intuitive and holistic manner of problem solving. A chess grandmaster, for instance, does not apply any rules to chess, as do beginners, but "sees" in a single glance the situation on the board, potential moves, and potential replies. The expertise consists not in a warehouse of facts and rules, but in the recollection of past situations that were successfully confronted. The simple rules that are taught beginners—such as "first the knights, then the bishops," or "a rook is usually worth more than a bishop"—are used as rules of thumb in the global context of thousands of perceived situations, plans, moves, and countermoves.

Rules of thumb are an important learning tool for the novice and advanced beginner in a particular knowledge domain, providing a simplified vision of its structure and a handle for addressing specific situations. Because reality lacks a formal structure that can be grasped in rules (contra the ontological assumption), expertise ultimately consists in a knowledge of and ability to deal with countless separate situations. Expert systems, which are based on the assumption that the knowledge of experts can be formalized, can never reach this level of sheer expertise.

This gives rise to a number of limitations in the range of application of (symbolic) expert systems. Because expert systems cannot make decisions or form judgments at the level of an expert, they cannot be entrusted with tasks that require expertise. However, Dreyfus is convinced that expert systems can often attain a certain degree of *competence,* a performance level that surpasses that of a novice or advanced beginner and is comparable to that of an advanced student. Expert systems therefore might indeed prove useful in applications that do not call for performance at the expert level.

One question that Dreyfus does not broach, however, is that of deciding whether a particular task calls for expertise or only competence. The determination of the right punishment for a crime clearly calls for the expertise of a judge, who takes into account the circumstances of the crime and background of the lawbreaker in establishing the punishment. A legislative body, however, may decide that judges (or juries) henceforth must decide the punishment based on a certain list of formal principles, such as the type of crime, the criminal's record, and a set of other verifiable data. This would eliminate the role of the intuitive judgment of the judge and transform the judge's function into the application of a number of formal rules—a task that could well be taken over by a competent expert system.

Whether the use of expert systems in particular domains can be justified thus largely depends on the legitimacy of the decision to formalize these domains and on the decision to exclude the role of intuitive judgment. Already in 1976, the AI researcher and critic J. Weizenbaum wrote an influential "critique of instrumental reason," which attacked the tendency to reduce human problems to calculable, logical problems. This phenomenon, of course, predated computers, but their use greatly stimulated the desire to make such attempts. Weizenbaum's conclusion, to which Dreyfus would clearly subscribe, is that the intuitive judgment of human beings is indispensable even in specialized domains.[3]

Besides expert systems, a second type of intelligent computer system that Dreyfus discusses (and that is closely allied with expert systems) consists of intelligent tutoring systems, or ITSes, which are employed in computer-aided instruction. Intelligent tutoring systems are computer programs that take over certain teaching roles. For the most part they are not intended actually to replace the teacher, but rather to supplement the instruction. An important distinction must be drawn between the use of a computer as an ITS and its use in other functions such as word processing, electronic whiteboards, or databases that use "unintelligent" computer programs. An ITS, by contrast, is a program that pretends to be intelligent, for it pretends to possess some of the abilities of a professional teacher.

Intelligent tutoring systems can help students in two ways. In its most simple form, an intelligent tutoring system can supply problems to which the student must find the correct answer—exercises in spelling or algebra, for instance. This kind of ITS has the ability to generate new questions or problems and to evaluate the answers given by the students—and through the repeated production of examples and exercises to help a student obtain knowledge and ability in a particular domain. Dreyfus has little problem with this kind of ITS, viewing it as a superlative application of the ability of computers to foster learning. The only danger with this application is that it works so well as to create the temptation to overuse it in the learning process, at the cost of other ways of learning.

A more advanced type of ITS takes a more active and participatory role in providing advice and instructions, in explaining what the student did wrong, and in selecting the problems and setting the pace for the individual student. This type of ITS is used in teaching complex knowledge and abilities—where the task, for instance, is to master certain theories and concepts and apply them to concrete situations. In order to do this, the ITS must have at its disposal a certain amount of didactic proficiency.

A first objection to this kind of ITS is that it is unsuited for helping students develop genuine expertise in a particular domain, for in order to do this the computer system itself needs to possess expertise. But as argued above, it is impossible to give expertise to computers that have been programmed according to symbolic AI. According to Dreyfus, intelligent tutoring systems are well suited to teach a certain measure of competence in an area. They are especially well suited to teaching the early stages of learning, in which the acquisition of rules still plays a major role. But it would be disastrous if such systems were used in later stages of learning, for they use only rules. They would block the acquisition of expertise, which requires at a certain point giving up the use of rules.

Even when ITSes are used only with novices and apprentices, another more serious problem arises. In order to teach well, an ITS must possess not only a great store of specialized knowledge, but also the ability to connect that with the knowledge the student already possesses, suitably adapting it in the process. A teacher of natural science, for instance, needs to have insight into the naive conceptions about the workings of nature that the students bring into class, and the ability to modify these to make way for the more advanced conceptions. Any ITS would have to possess similar insight and ability.

But the problem is that an ITS is able to express its knowledge and ability only in terms of a number of symbols and rules. The ITS assumes implicitly that the student is a rational, symbol-manipulating, rule-following being. In fact, however, the student is an embodied being that dwells in a human world, and the ITS needs to be able to put itself in the student's place in order to understand where the student is coming from. Because it is unable to do this, an ITS will be unable to help students in seeing the underlying connections that will enable them to master a new knowledge domain. In conclusion, the problem with ITSes used as full-blown teachers is that they cannot help advanced students because they do not themselves possess expertise in the relevant knowledge domain, and they cannot adequately help novices and advanced beginners because their didactic skills are lacking.

Dreyfus concludes that existing intelligent computer systems, especially expert systems and intelligent tutoring systems, foster the impression that the human mind works like a computer. They promote an ultimately erroneous conception of knowledge as something that can be formulated in explicit rules and principles. In the process, the intuitive ability and expertise of human beings, which cannot be grasped through formal rules, becomes devalued, and students are encouraged to seek knowledge and skills according to the rationalistic model. Eventually this may alter the self-image of human beings to the point where they will begin describing themselves in rationalistic terms as abstract thinking machines. This is the tendency that Dreyfus fears and wants to change.[4]

THE VALIDITY AND INFLUENCE OF DREYFUS'S WORK

Already in 1965, Dreyfus prophesied that symbolic AI would end up failing to achieve a full and complete imitation of human intelligence. Over the years he has systematically criticized the predictions and expectations that have been projected onto new projects and approaches of symbolic AI. And in many respects Dreyfus has been proven right. Although symbolic AI has certainly scored a number of successes, the results in many areas have been disappointing. Thus no computer programs have yet been developed that can understand natural language well and answer open-ended questions about a text, that can interpret the meaning of images, that can allow a robot to navigate successfully in a messy environment, and that can solve creative problems. Dreyfus's critique of neural network AI is harder to evaluate, for the field is still too young, though the problem of generalization that he describes remains unsolved.

Two recent projects in symbolic AI are worth considering, as they have the potential to undermine Dreyfus's position. The CYC project headed by AI researcher Douglas Lenat had as its goal to develop a knowledge base of over one million assertions or "rules" that would codify most of the general background knowledge necessary for computer systems to intelligently engage in natural language communication: common-sense knowledge, including such facts as "People normally wear underwear" and "If an object is not supported by another object or surface, it will fall down." CYC was begun in 1984 as a ten-year project. If successful, it would disprove that the common-sense knowledge problem of symbolic AI is insurmountable, and that common sense can be programmed into a computer, with enough effort. The completed knowledge base of CYC is currently being marketed for application to various tasks, such as intelligent database retrieval, and improved machine translation and speech recognition. Indeed, applications based on CYC's knowledge base may perform better in these and other areas than systems that lack its extensive knowledge base. However, CYC is no longer expected to *solve* the problem of machine translation, or other major problems in natural language understanding, which is what one would expect if CYC's common sense matched human common sense. So the CYC project has not so far refuted the validity of the common-sense knowledge problem of symbolic AI.[5]

The victory of chess computer Deep Blue over world champion Garry Kasparov in their 1997 match seems to strike a more powerful blow against Dreyfus's position. Deep Blue is a system built upon the principles of symbolic AI. Dreyfus's position that symbolic expert systems cannot perform at the expert level seems to be refuted by the performance of Deep Blue.

Dreyfus has always made an exception for expert systems operating in domains of knowledge that are completely formalizable, such as some domains of mathematics and formal logic. Yet, whereas the rules of play of chess are fully formalizable, the best move in a chess game cannot be calculated, as it can be in a game like tic-tac-toe. This is because chess games can potentially drag on forever, and hence there is not a finite number of moves to consider that can be evaluated as part of a strategy for winning the game. Thus, chess computers have to resort to heuristic rules that anticipate the myriad possible strategies of the opponent.

Still, there are important differences between the knowledge domain of chess and other domains of expert knowledge that make it unlikely that the success of Deep Blue can be generalized. This is because the domain of chess can be made to *resemble* a completely formal domain that contains a finite search space in which an optimal solution can be found. If it is taken into account that most chess games are less than one hundred moves, then for practical purposes, the number of possible moves in a chess game is finite, although still too vast to have all possibilities considered. The challenge for chess computers is then to perform calculations only in the more promising parts of this search space. Deep Blue was able to do this with the aid of a vast database containing former chess games played by Kasparov that allowed it to direct its calculations to that part of the search space that Kasparov had occupied in past games. It is unlikely that this strategy of targeted calculation can be applied to other domains of expertise. The problem in other domains, like medicine or economics, is that they cannot be made to resemble a finite, formalized search space to which rules can then be applied, because the elements and regularities in these domains are not formal to begin with. Deep Blue's success hence does not undermine Dreyfus's general position on expert systems.

Not only has Dreyfus been proven right in many of his predictions on the success of AI, but AI research has moved more and more in the direction of Dreyfus's alternative theory of intelligence. This is true, for instance, of the emergence of neural network AI, which as Dreyfus points out is fully compatible with his own ideas about intelligence. It is also true of the well-known work of Agre and Chapman at MIT (Agre 1988; Chapman 1991), which is sometimes called "Heideggerian AI" because it tries to implement in AI a number of views promoted by Heidegger and Dreyfus, such as that intelligence is situated in a world and does not require rules, and that actions can be goal oriented in the absence of explicitly represented goals.

The situatedness of intelligence is also a central point of departure of the work of the noted AI researcher Terry Winograd and his colleague Fernando Flores. They want to base not only AI research, but also the design of other computer systems, on Heideggerian principles.[6] Winograd and Flores argue that the design of computer systems as well as its internal logic need to take into account and reflect the fact that these systems must function in a human world and communicate with human users. Computers must be prevented from imposing their own rationalistic logic on the surroundings in which they function.

Even the idea that intelligence presupposes possession of a body has struck a responsive chord in AI research. A recent project at MIT that has drawn much international attention, for instance, is the Cog-project under the direction of Rodney Brooks. The principal assumption of this project is that human intelligence requires human interactions with the world, and therefore a body in which such interactions are possible (Brooks and Stein 1994). Cog is a robot that is equipped with artificial sensory organs (including sensors that keep track of the position of its own body), a voice, and steerable limbs. Cog's "mind" consists of a computer system that is to some extent distributed throughout his body. The aim is to have Cog acquire sensorimotor intelligence thanks to its sensorimotor interactions with the environment, and perhaps develop "higher" forms of intelligence on top of these more basic abilities.[7]

Much of the inspiration for the development of such work can be traced back to the work of Dreyfus himself. Dreyfus was the one who introduced the ideas of thinkers like Heidegger and Merleau-Ponty into the AI world. The work of such AI researchers as Winograd and Flores, and Agre and Chapman, was explicitly inspired by his ideas. Many other AI researchers, even including followers of symbolic AI like Minsky and John McCarthy, admit that Dreyfus's critiques have influenced their own research. Dreyfus is living proof that philosophers can indeed play a major role as critics of, and commentators on, science and technology in practice.

NOTES

1. The assumptions of classical AI on the nature of intelligence are not only found within AI, but gave rise in the seventies and eighties to a new, interdisciplinary science

called cognitive science. Cognitive science is the science of both biology-based and artificial intelligent processes, and has emerged as a result of collaborations between AI researchers, psychologists, philosophers, and linguists. Nowadays only part of the research in cognitive science is based on the assumptions of classical AI, and other research traditions have developed as well, such as the paradigm of neural networks.

2. For an introduction to neural networks, see Anderson, *An Introduction to Neural Networks*.

3. A more recent and exhaustive critique of expert systems is found in Collins, *Artificial Experts*. For ethical discussions about expert systems, see Forester and Morrison, *Computer Ethics*, ch. 7.

4. The danger foreseen by Dreyfus that human beings will begin to see themselves as computers has to some extent already come true, as indicated by the psychological studies in Sherry Turkle's *The Second Self*.

5. See Dreyfus, *What Computers Still Can't Do*, for a theoretical critique of the CYC project.

6. See for instance Winograd and Flores, *Understanding Computers and Cognition*, and Winograd, "Heidegger and the Design of Computer Systems"; also Suchman's influential *Plans and Situated Actions*.

7. The notion of intelligence as a situated and embodied phenomenon has also gained ground in psychology and cognitive science. See, e.g., Clark, *Being There;* Varela, Thompson, and Rosch, *The Embodied Mind;* Johnson, *The Body in the Mind;* Lakoff, *Women, Fire and Dangerous Things*.

BIBLIOGRAPHY

Agre, Philip. 1988. *The Dynamic Structure of Everyday Life*. Cambridge: MIT AI Lab Technical Report 1085.

Anderson, James. 1995. *An Introduction to Neural Networks*. Cambridge: MIT Press.

Brooks, Rodney, and Lynn Andrea Stein. 1994. "Building Brains for Bodies." *Autonomous Robots* 1: 7–25.

Chapman, David. 1991. *Vision, Instruction, and Action*. Cambridge: MIT Press.

Clark, Andy. (1996). *Being There: Putting Brain, Body, and World Together Again*. Cambridge: MIT Press.

Collins, Harry. M. 1990. *Artificial Experts: Social Knowledge and Intelligent Machines*. Cambridge: MIT Press.

Crevier, Daniel. 1993. *AI: The Tumultuous History of the Search for Artificial Intelligence*. New York: Basic Books.

Dreyfus, Hubert L. 1965. *Alchemy and Artificial Intelligence*. The RAND Corporation Paper P-3244.

———. 1967. "Why Computers Must Have Bodies in Order to be Intelligent." *Review of Metaphysics* 21: 13–32.

———. 1972. *What Computers Can't Do: A Critique of Artificial Reason*. Cambridge: MIT Press.

———. 1991. *Being-in-the-World: A Commentary on Heidegger's "Being and Time."* Cambridge: MIT Press.

————. 1992. *What Computers Still Can't Do: A Critique of Artificial Reason.* Cambridge: MIT Press.

————. 1996. "Response to my Critics." *Artificial Intelligence* 80: 171–91.

Dreyfus, Hubert, and S. E. Dreyfus. 1986. *Mind over Machine: The Power of Human Intuition and Expertise in the Era of the Computer.* New York: Free Press.

————. 1988. "Making a Mind versus Modeling the Brain: Artificial Intelligence Back at a Branchpoint." *Daedalus* 117: 15–43.

Forester, Tom, and Perry Morrison. 1994. *Computer Ethics: Cautionary Tales and Ethical Dilemmas in Computing.* Cambridge: MIT Press.

Hoven, M. Jeroen van den. 1995. *Information Technology and Moral Philosophy.* Ph.D. thesis, Erasmus Universiteit, Rotterdam.

Johnson, Mark. 1987. *The Body in the Mind: The Bodily Basis of Meaning, Imagination, and Reason.* Chicago: University of Chicago Press.

Lakoff, George. 1987. *Women, Fire, and Dangerous Things: What Categories Reveal about the Mind.* Chicago: University of Chicago Press.

Suchman, Lucille. 1987. *Plans and Situated Actions: The Problem of Human-Machine Communication.* Cambridge: Cambridge University Press.

Turkle, Sherry. 1984. *The Second Self: Computers and the Human Spirit.* New York: Simon & Schuster.

Varela, Francisco, Evan Thompson, and Eleanor Rosch. 1991. *The Embodied Mind: Cognitive Science and Human Experience.* Cambridge: MIT Press.

Weizenbaum, Joseph. 1976. *Computer Power and Human Reason.* San Francisco: W. H. Freeman.

Winograd, Terry. 1995. "Heidegger and the Design of Computer Systems." In *Technology and the Politics of Knowledge,* edited by A. Feenberg and A. Hannay. Bloomington: Indiana University Press.

Winograd, Terry, and Fernando Flores. 1986. *Understanding Computers and Cognition: A New Foundation for Design.* Norwood, N.J.: Ablex.

Wrathall, Mark, and Jeff Malpas, eds. 2000. *Heidegger, Authenticity, and Modernity: Essays in Honor of Hubert L. Dreyfus, Vol 1.* Cambridge: MIT Press.

————. 2000. *Heidegger, Coping, and Cognitive Science: Essays in Honor of Hubert L. Dreyfus, Vol. 2.* Cambridge: MIT Press.

3. Andrew Feenberg: Farewell to Dystopia

Hans Achterhuis

The transition from the earlier austere and monolithic classical philosophy of technology to a more empirically attuned and nuanced version is nowhere clearer or sharper, among the thinkers represented in this volume, than in the works of Andrew Feenberg. Feenberg, a professor at San Diego State University, began his career as a faithful disciple of the neo-marxist school of the critique of technology, but has developed into an original and versatile thinker concerned with the ambivalent possibilities of technological culture. He has never repudiated his original inspiration in critical neo-marxism. Instead, he has appropriated new developments in the philosophy of technology, including social constructivism, in order to concretize and actualize his otherwise rather abstract and naive neo-marxist perspective. The question of whether its form and content have altered so much in the process that Feenberg cannot still legitimately appeal to it is one to which I shall return at the end of this chapter.

Feenberg's first book (1981), *Lukacs, Marx, and the Sources of Critical Theory,* is a meticulous analysis of two works that are highly prized by leftist intellectuals: Lukacs's *History and Class Consciousness* and Marx's *Economic and Philosophical Manuscripts of 1844.* As hinted at by the title, Feenberg finds in these two works the roots of critical theory; that is, of the thinking of the Frankfurt School, whose most important representatives are Adorno, Horkheimer, and Marcuse. This book is of little relevance in an introduction to Feenberg's philosophy of technology. For one thing, technology plays only a minor role in it; the few citations of "technology" and "technique" in the index refer to passages in which these concepts are mentioned in passing in discussions of other topics, such as Lukacs's theory of the reification of the social relations in capitalist society. Furthermore, and more generally, a contemporary reader will not find a close reading of *Lukacs, Marx, and the Sources of Critical Theory* especially rewarding. Many passages practice the kind of fastidious exegesis of sacred texts and indulgence in polemics with other interpreters who are deemed to be insufficiently orthodox that was popular some decades ago but has not worn well. These party-line analyses—and the few remarks about technology found in them—are not especially relevant to current debates.

Feenberg's second book, *Critical Theory of Technology* (1991), retrieves a number of the central positions of the first. As the title indicates, it is unlike the previous work in that it is directly concerned with technology. But in many respects this book is a hybrid. Abstract and even highly strained observations drawn straight from orthodox marxism are interwoven with more current— and legitimately philosophical and sociological—analyses of technological developments. The persistence of a rigid (neo)marxist framework, derived as much from Marx as the Frankfurt School, makes it difficult to fully appreciate the very interesting ideas of Feenberg himself. His entire argument is based on the assumption that it is possible and desirable to make a transition from the present capitalist society to a socialism conceived as a radically other and better society—a binary scheme that obliterates any difference between different concrete technological developments and possibilities inside capitalism. Concrete improvements are dismissed as contemptible reforms. What alone counts for Feenberg are steps that break radically from the capitalist logic and technological code en route to this fundamentally new society. However, until this society is concretely in place—and despite much diligent research Feenberg manages to turn up only the vaguest hints in recent marxist literature of what it might be like —Feenberg finds that the world will only become steadfastly more uniform through the march of modern technology.

In his third book, *Alternative Modernity* (1995), Feenberg for the most part leaves behind this absolute opposition between the corrupt present and the desirable future good. What takes its place is a sharp and convincing analysis of a plurality of technological developments and possibilities. Feenberg now recognizes in world technological culture very different specific technological cultures; the socialist alternative to the current, restricted capitalist technological form of modernity is fanning out into a multiplicity of alternative modernities. Typical of the differences between this book and Feenberg's previous one is the vision of Japan expressed in each. In *Critical Theory of Technology* (11–12), Feenberg makes only passing reference to Japan, lamenting its failure to create a culturally original form of modern society. Confronted with "technology's universalizing pressures," Feenberg writes, Japanese culture was shorn of its uniqueness until it came to differ only very superficially from that of the West. In *Critical Theory of Technology* the only trace of an alternative (socialist) modernity that Feenberg can discern—consistently enough, given his dualistic framework—is the Yugoslavian workers councils and certain experiments carried out during the Chinese Cultural Revolution (133). But in *Alternative Modernity,* Japan figures more prominently and positively, as an example of a technological culture that has more or less succeeded in giving shape to an original, alternative form of modern society.

Another significant difference between the two studies of the philosophy of technology lies in their relation to empirical reality. This relation is absent, often in blatant ways, from *Critical Theory of Technology.* In chapter 3, for instance, Feenberg considers the problems of the Soviet transition to socialism only via an analysis of Lenin's conceptual confusions in *The State and Revolution;* Lenin's practical politics, not to mention those of Stalin, are barely discussed. Feenberg even avoids analyzing the concepts and especially the practical developments of a movement like anarchism, which could have added much to his discussions. For Feenberg, only the orthodox marxist texts seem to count. This disregard for the diversity of reality we also see in his treatment of actual technological developments. An otherwise stimulating chapter on computers and artificial intelligence remains for the most part stuck in theoretical discussions that contrast—once again!—two different visions of the introduction of the computer, and point to two totally different ways (emancipation and oppression) in which technology can evolve. This rigid and abstract framework is shattered in the chapter on computers in *Alternative Modernity.* In that book, too, Feenberg's theoretical discussions are often oriented to empirical issues, as illustrated by his extensive and convincing discussion of what happened in France in the wake of the introduction of the so-called Minitel, a terminal, distributed free by the French telephone company, that was designed to look like an accessory to the telephone. The entire book is, in fact, organized so that each theoretical chapter is followed by one with an empirically oriented concretization.

This chapter on Feenberg's philosophy of technology will focus principally on *Alternative Modernity,* making some use of *Critical Theory of Technology* and his latest book *Questioning Technology.* The aim is not to provide a critique, but rather to point out the characteristic features of Feenberg's thought. I shall proceed as follows: A brief exposition of the Frankfurt School's philosophy of technology will be followed by a discussion of the way Feenberg concretizes and actualizes it in and through his interpretation of the social constructivist appropriation of technique. Then the dystopian vision of technological culture that we encountered in the critical theorists will be critiqued partly along the same social constructivist lines. An analysis of the novel *The Master of Go,* by Japanese Nobel laureate Yasunari Kawabata, offers at this point a fascinating entry into the question of the possibility of an alternative modernity. Finally, this all-too-brief overview of the work of a prolific innovator in the philosophy of technology examines the concrete alternative to classical philosophy of technology that Feenberg offers in his most recent book, *Questioning Technology.*

THE DEBATE BETWEEN MARCUSE AND HABERMAS

The debate between Marcuse and Habermas on the place of technology in critical social theory plays an important role in all three of Feenberg's books on the philosophy of technology, though his assessment of this extremely important philosophical debate is different in the first and later books. An outline of the main points of the debate, and of Feenberg's creative reappropriation of it will have to suffice.

In the introductory chapter of *Critical Theory of Technology,* Feenberg distinguishes between his critical theory and an instrumental theory on the one hand and a substantive theory on the other (Feenberg borrows this division from Borgmann 1984, 89). The instrumental theory, which is rarely found now among philosophers of technology, is still the prevailing view in social debates about technology. Its central assumption is that technology is purely instrumental, utterly neutral with respect to political and social choices. This social-political neutrality is said to result from the rational and universal character of technology. Modern technology functions always and everywhere in the same way, and efficiency is its only true norm. The substantive theory, on the other hand—whose leading expositors are the classical thinkers of technology such as Heidegger and Ellul—questions all those characteristics that the instrumental theory ascribes to technology: instrumentality, neutrality, rationality, and universality. The substantive theory claims that technology amounts to a new cultural system that draws the whole of natural and social reality under its logic of domination and that sweeps away all traditional relations to the world. Under its sway, human beings as well as nature are reduced to interchangeable and calculable parts of a totalitarian technological system.

Feenberg highlights the differences between the two approaches using the example, dear to Borgmann and Ihde, of what happens at mealtime. The replacement of the traditional family meal by the individual consumption of fast food illustrates the unintended but real consequences of the technological approach to food consumption. To consider food only as the vehicle for the ingestion of a necessary amount of calories and nutrition, to be consumed as quickly and efficiently as possible, is to misconstrue the extensive cultural changes that go along with this diminished conception of what a mealtime is. Just how far reaching these cultural changes might become in the future is suggested by the results of a workshop that a student of mine participated in of a Dutch interdisciplinary research program on sustainable technological development. This workshop focused on the far future (it aimed to look backward from the year 2040), and sought to imagine the development of extremely environment-efficient technology that would reduce human spatial requirements on the earth's physical environment by a factor of 20. One of the imagined developments was the existence of a pill able to supply all the nutritional requirements prescribed by an individual's "genetic card," taking into account not only one's individual needs but also personal tastes. Clearly, such a food pill would be more than a neutral, rational nutritional instrument; it would bring sweeping cultural changes.

But despite the obvious and substantial differences between the instrumental and substantive views of technology, Feenberg perceives an important affinity: both are expressions of what he calls a "take it or leave it" attitude. Technology is either embraced in its quasi-totality for its instrumental character or repudiated because of the cultural and social threats it poses. And when either approach does succeed in seriously discussing the political and social applications of technology, the discussion is primarily couched in terms of the proper controls and restraints; proposals for the social regulation of technological developments mainly involve setting moral and political limits. When advocates of the instrumental approach see that the advancing efficiency of technological instrumentation threatens values, habitats, or natural preserves, they are often prepared to adapt, willing to compromise efficiency in order to preserve certain essential values. Advocates of the substantive approach tend to be better able to set the necessary limits more sharply and rigorously. They find that a more "natural" way of living may require significantly slowing down or even rolling back technological developments in areas such as biotechnology.

In Feenberg's analysis, both approaches fail to address the content of technology in their discussions, and also fail to address the possibility of an alternative technology. Both of these aspects, however, are central to critical theory as developed by members of the Frankfurt School. That theory shares with the substantive theory the idea that modern technology is a cultural system whose influences are oppressive. Critical theorists who discuss the reification of social relations or the one-dimensionality of society, for instance, tend to highlight just this substantive dimension of technology. In contrast with Heidegger's ahistorical approach, which sees technological domination as the specific constellation of 'Being' in the modern era, the thinkers of the Frankfurt School connect modern technological domination with the emergence of capitalist consumer society. Under capitalism all qualitative values and ends are abstracted, and only the quantity of calculable means counts. The broad rationality of the Enlightenment that aimed to foster an inclusive discussion about values and objectives has been winnowed down to a debate about instrumental rationality that is interested only in the efficiency of various means. A calculus of utility of the sort that utilitarians preach has replaced the discussion of the good life and the just society.

If modern technology is indeed the expression of a particular historical constellation of circumstances governed by a diminished type of rationality, it implies in principle at least the possibility of another technology grounded in a broader type of rationality. And if modern technology is less a metaphysical event than the result of the actions of a specific ruling class, it implies that human beings can change it. Marx (who in many respects had an instrumental approach to technology) assigned to the oppressed working class, the proletariat, the responsibility for effecting change; Feenberg develops an alternative marxist interpretation in which the content of technology would have to change substantially following a socialist revolution. But by the middle of the last century the Frankfurt School had ceased to await an approaching socialist revolution, and ceased to view the proletariat as the revolutionary subject. And existing forms of socialism in Eastern Europe and Asia failed to provide inspiring visions for how technological developments could be given a substantially different direction and content. Lacking concrete possibilities and examples, Horkheimer and Adorno gave themselves over to vague utopian images of a possible future reconciliation of human beings and nature.

Marcuse, however, went a step further, and sought to put the issue of a substantially different technology on the contemporary philosophical and social agenda, though to be sure in a very abstract way. He dared speculate about the end of "scientific rationality in its established structure and direction" (Marcuse 1964, 230–31). For Marcuse, any further technological advances could and should involve a break in which quantity again changes back into quality. The scientific-technological program of a liberated socialist society would be "open to objectives which go beyond utility." Instead of domination it would aim at the "art of living." The new social ends could be internalized in technological development itself, says Marcuse; "as technical ends, they could then operate in the project and construction of the machinery" (232). Technology's brute exploitation of nature could come to an end in this possible future "pacified existence." Marcuse foresaw nature as becoming "liberated" through a technical remodeling in which it is treated as a partner rather than an enemy of humans. Aesthetics and eros could then take the place of repression and exploitation.

Habermas challenged this idea in a famous article, "Technology and Science as Ideology," written in 1968 on the occasion of Marcuse's seventieth birthday, in which he attacked the direct relation Marcuse drew between (capitalist) domination and technology. Habermas saw technology as linked with the interests of all humanity rather than those of a specific class. He distinguished between work and interaction as two essential forms of human activity. The first concerns the dealings that human beings have with nature, the second social, interpersonal relations. The first is limited to purposive, goal-driven rational action, while the second involves symbolic interaction, mediated by norms and values.

According to Habermas, in the sphere of work it is impossible to conceive of a different technological interaction with nature. He claims that technological evolution follows a definite and unalterable direction, and underscores the point with a well-known image borrowed from Arnold Gehlen (Habermas 1968, 56): "At first the functions of the motor apparatus (hands and feet) were augmented and replaced, followed by energy production (of the human body), the functions of the sensory apparatus (eyes, ears and skin), and finally by the functions of the governing center (the brain)." Whoever dreams of different technological possibilities, according to Habermas, is a romantic in the worse sense of the word, and he dismisses the "secret hope" for a resurrection of nature that he finds in the imagery of Benjamin, Horkheimer, and Adorno, as well as Marcuse's discursive approach to the possibility of an alternative technology. For Habermas, any technology other than that of the present is unimaginable.

Habermas argues that instead of dreaming about a completely different technology, marxist thinkers should instead focus their attention on the sphere of social interaction, where domination and exploitation exist and the means for liberation and emancipation must be found. Marcuse's intuition is correct that the course of late capitalism has altered the nature and forms of social conflict—but this has less to do with the internal structures of science and technology than with the ways in which these legitimate domination and exploitation. The enormous surpluses generated by modern technology, for instance, have decisively changed the consciousness of oppressed social groups. Habermas sees the greatest dangers looming wherever science and technology threaten to overpower the sphere of interaction; in a technocracy, for instance, there would be no room left for the formation of social judgments about norms and values. Therefore, the intrinsic character of the sphere of interaction, Habermas feels, must be protected most of all against the imperialism of technology. In his later work, Habermas appeals to the well-known distinction between system and lifeworld. Each has its own intrinsic values that must not be allowed to interfere with each other. The great danger faced by the lifeworld is the threat of its colonization by the system, which must be constrained within appropriate limits. Within these limits, however, the system obeys a lawful necessity that in turn must not be contaminated by the norms and values of the lifeworld. It is precisely an achievement of modern societies, as compared with traditional ones, that the system has freed itself from the context of worldviews colored by myth and religion.

MARCUSE IS PARTIALLY RIGHT

Feenberg is doubtless correct that Habermas won the philosophical debate with Marcuse about the role of technology. This has partly to do with Marcuse's inability to concretize his ideas about the possibility of an alternative technology; whenever he tried to provide an illustration of a possible alternative technological engagement with nature, these were unconvincing. The "liberating transformation" of nature in "gardens and parks, and reservations" (Marcuse 1964, 240) is an appropriate ambition for an agricultural technology unencumbered by heavy demands on labor, space, or production but not a realistic ambition for contemporary food production. But Feenberg also claims that Habermas's victory was partly due to the times, for as the revolutionary fervor of the 1960s waned, more sober, concrete, and pragmatic approaches gained the upper hand over radical and idealistic abstractions.

Feenberg sees the outcome of this debate as disastrous for the philosophy of technology. Habermas's many fruitful ideas only conceal the fact that in many respects he falls back on an instrumental view of technology. For him it is unnecessary either from the social or the philosophical perspective to reflect on the content and development of technology; this is to be left to the engineers and other experts. What needs to be investigated, Habermas claims, are the social consequences that attend the introduction of specific technologies, in order to discover their proper roles and to set their boundaries.

Against Habermas, Feenberg wants to raise the question of the content and direction of technological developments within the philosophy of technology. And with Marcuse he believes that an alternative technology is possible, and that social values can even decisively change the process of technological development. Feenberg finds it possible to reprise these most important theses of Marcuse thanks to the many new ideas that have emerged since the 1960s in studies of technology, one of the most important of which is the idea that a technological artifact is a social construction; that is, it cannot be understood solely through technological norms of efficiency, for social norms play an extremely important role in its development and use.

In his first two studies on technology, Feenberg makes use of the work of French sociologist Bruno Latour to illustrate the new insights into the processes of technological development. Feenberg discusses Latour's work cautiously and somewhat ambivalently in *Critical Theory of Technology;* in *Alternative Modernity* the French thinker is assigned a more prominent and unwaveringly positive role. I shall discuss both interpretations in turn.

In *Critical Theory of Technology* (81–82), Feenberg uses certain concepts borrowed from Latour to help elaborate an idea, partly borrowed from Marcuse, of a technological code, distinguishing in a freely abstract way a capitalist from a socialist code. In the technological code, technological demands such as efficiency are inextricably bound up with the prevailing norms and values. More exactly, technological norms are not context-independent, as often asserted; they are often established through much social influence and conflict. In retrospect, the engineer does not see the original social determination of these supposedly "internal" norms. Feenberg's best and most persuasive example of this appears in an early article (1982) and concerns the abolition of child labor in the nineteenth century. Initially the abolition aroused considerable opposition based on economic as well as technological considerations. Society, claimed the experts, should not interfere with processes that are highly efficient because they are well suited to being performed by children, and the loss of productivity that would occur if child labor were to be abolished would have disastrous consequences for the economy. This did not, in fact, happen; productivity ultimately increased rather than decreased, though certain modifications in the technology had to take place for this to occur. Today, Feenberg points out, no one would seriously consider on technical grounds using child labor to boost the efficient design of machines and work environments. The social norms that prohibit child labor have been completely internalized in the designing process, and the initial protests by the experts who objected that social criteria should not interfere with technological requirements have been forgotten.

Similarly, Feenberg shows, safety features have come to be integrated into the design of many construction processes, often after fierce social battles in which these features were opposed as interfering with the beneficial influence of "pure" technological criteria. Once again, Latour provides the theoretical underpinnings for Feenberg's examples, as for instance in the analytical distinction between *sociogram* and *technogram* (Latour 1987, 138–40). For each socially accepted technological object (and Latour describes taken-for-granted technological artifacts and scientific facts as "black boxes"), one can outline a sociogram and a technogram. The first refers to the external alliances of social interests that are indispensably involved in shaping the social role of the artifact and in its social acceptance, while the second refers to the internal alliances that are involved with the technical elements of the artifact. Latour's point is that the technogram does not happen first, followed by the sociogram; rather, the processes are intertwined.

Sociogram and technogram are two sides of the same technological artifact or system, whose success is as much a function of its particular configuration of technological elements as of the specific relationships of a network of users. Latour's own examples include the development of the diesel engine and of the serum for diphtheria inoculations into "black boxes," socially accepted technologies. Both developments, Latour shows, can be understood only if we grasp the laws of the ongoing reciprocal relationship between sociogram and technogram. Potential users were recruited at each technological step in the laboratory or factory, and their needs or desires were incorporated each time into the technology. Thus only an analytical distinction can be made between sociogram and technogram; in social practice they are inextricably intertwined in what Feenberg calls the "technological code."

In *Critical Theory of Technology* Feenberg uses Latour's ideas, counter to Latour's own expressedly antimodern orientation, to elucidate the difference between two mutually exclusive technological codes, the capitalist and the socialist. In *Alternative Modernity* Feenberg recognizes the inadequacy of this oversimplified polar distinction and, now much more in line with Latour, fragments it into many codes. Each black box, in fact, has its own technological code that can be brought to light only by examining its sociogram and technogram. To illustrate Feenberg's procedure, I shall discuss below how he handles the Minitel example.

First, however, I want to discuss another important concept related to the idea of the technological code that Feenberg, in *Alternative Modernity*, borrows from Latour. Not only are social norms integrated into the developmental process, but maintaining them is also *delegated* to things (Feenberg 1995, 83–85). Moral imperatives, for instance, are often not left to human beings to follow on their own, but are materialized in objects. One of Latour's examples involves seat belts (Latour 1989). The existence of the seat belt, the warning light if one does not buckle it, and the device that partly extends it over the occupant of a car as the door shuts admonish or force the occupant into safe driving practices. Feenberg uses an even more mundane example to illustrate the moral power of things. Whenever we lock up our private property, he says, we are substituting a strategic material intervention for a moral appeal. At the same time the lock presupposes a certain moral consensus, and the idea of the legitimacy of private property is reinforced. Material technological interventions thus not only change the world, but also instruct our moral behavior and tell us which values are important. In technological cultures they form an inseparable part of communicative processes that contribute to the formation of moral consensus. Against MacIntyre (1981), who laments the absence of moral debate from modern society, Feenberg claims with Latour that the moral debate is taking place in another domain, one that MacIntyre is overlooking and in which technological artifacts are key players—in the "parliament of things," as Feenberg puts it in the title of the first chapter of *Critical Theory of Technology.*

The final important concept that Feenberg borrows from Latour—and uses to concretize Marcuse's abstract ideas—is *underdetermination.* By this he means that there is seldom a single, decisive reason why one particular techno-logical design should be preferred to others on the basis of efficiency—that there is "one best way"—as Ellul had also claimed. Technological principles by themselves are usually insufficient to specify a single design. And as Marcuse showed, though in a very general and highly abstract way, social interests play a continual role in the development of technologies, and other options are also possible. "Technological choices are thus 'underdetermined,' and the final de-cision about alternatives ultimately depends on the 'fit' between them and the interests and beliefs of the various social groups that influence the design pro-cess" (Feenberg 1995, 4).

THE WAY OUT OF DYSTOPIA

Thanks to Marcuse's critical position, and to the new insights of technol-ogy studies, Feenberg has opened up an important field of research for a revised critical theory of technology. Before examining more of the moves he has made in this new terrain, it is important to note how he disposes of one obstacle hanging over many critiques of technology: the conception of modern techno-logical society as a dystopia. The original critical theory doubtless had many dystopian elements, as did the classical philosophy of technology. This dysto-pian heritage, which Feenberg also shared in his first book about technology, should be severed in order to understand the new phenomenon of techno-logical culture. But it has proven extremely difficult for philosophers of tech-nology to shake free of that dystopian legacy, and Feenberg remains caught up in it to some degree even in *Alternative Modernity.*

Already in one lengthy passage of *Critical Theory of Technology* (72–78), Feenberg checked off the paradoxes of dystopia. In *Alternative Modernity* this idea is central to the second and third chapters. In the relevant chapter of the former book Feenberg approvingly compares the theories of Marcuse and Foucault on the intertwining of technological-scientific and social rationality.

Marcuse's theory, as already mentioned, boils down to the view that "technological rationality has become political rationality" (Marcuse 1964, xvi). Technological domination and political domination are two sides of the same totalitarian late-industrial reality. Social domination is anchored in the technological
project, which Marcuse understands in the broad sense as a fundamental historical project—namely, "the experience, transformation, and organization of
nature as the mere stuff of domination" (xvi). Foucault means something similar when he asserts that knowledge and power are inseparably linked with and
mutually strengthen each other, a phenomenon that he saw clearly expressed
in the "disciplines" that appeared in the eighteenth century, social technologies designed to discipline and normalize the population. These technologies
often crystallized into material forms, in the construction and arrangement of
houses, hospitals, jails, schools, and cities. The most famous image of this is
Bentham's Panopticon, a jail architecture consisting of a tower around which
the cells were arranged in such a way that they could all be observed from a
central point, facilitating the control and disciplining of individuals. The
knowledge produced is not neutral, but always allied with the exercise of
power—not dependent on good or bad personal intentions or interests, but
integrated into the technological program. Foucault suggests, with his peerless
prose, that contemporary society is essentially one great panopticon (Foucault
1977).

For Feenberg, the problem starts here. Both Marcuse's image of a one-
dimensional world and Foucault's of a panopticon characterize sealed-off
worlds from which escape is impossible. These worlds are dystopias—Marcuse's is often reminiscent of Huxley's *Brave New World* and Foucault's of
Orwell's *1984*—and their subjects, thoroughly permeated and controlled by
power relations, lack any foothold for oppositional action. Feenberg notes correctly that both thinkers do call for such action, but also analyzes at great
length the theoretical impasse that results, in which the calls of both thinkers
for liberation and oppositional activity ring false. Feenberg connects this impasse with their failure to develop anything more than a superficial philosophy
of technology; for both, technology is ultimately conceived in an oversimplified, facile manner as technological domination. Feenberg claims to overcome
the dystopian and apocalyptic thinking of both Marcuse and Foucault through
his idea of the technological code, which brings to light the specific hegemonic
role that contemporary technology plays in maintaining the values of the ruling
class and that could be replaced by another, socialist technology, while preserv-
ing their most important insights concerning the social and political ladenness
of technology.

Feenberg, in fact, finds that virtually the entire tradition of classical philosophy of technology, including Heidegger, Ellul, and Illich in different ways,

has fallen victim to dystopian thinking. All of these thinkers share the idea that we are delivered over to technology, that our individual existence and ways of life have been desiccated through the domination of a great technological system. Everything to which we owed our humanity has eroded away. Our technological instruments and systems more and more determine who we are. In the shadow of a cathedral many different kinds of traditions could flourish, they claim, while superhighways or atomic power networks damage such traditions in irreparable ways. Feenberg sees but little difference between these thinkers and Marcuse and Foucault. While Marcuse and Foucault call, however illogically, for political opposition, Heidegger, Ellul, and Illich call instead for an attitude of personal resignation. Nevertheless, all five thinkers share the view that technological culture is a dystopia from which escape is impossible.

In an interesting chapter on "Dystopia and Apocalypse," Feenberg examines the lingering persistence of a critical attitude toward technology as it has appeared in post-World War II popular culture, such as in science fiction, novels, movies, television, and advertising. This attitude reaches epic proportions in James Bond books and movies: our hero, sent into battle equipped with a dazzling array of technological gadgets, ultimately triumphs over evil thanks to his own human power and individual resourcefulness; meanwhile, the villain is always portrayed as a technocratic maniac who rules his accomplices with an iron grip. The public fears of a technological dystopia are often, Feenberg notes, strangely enough projected onto villains who are based in some exotic location in the Third World—but the evils of technological domination were soon also associated with Western social life as well. A better image of the triumph of technology over traditional culture can hardly be found than in the striking scene that opens the Fellini film *La Dolce Vita:* a giant statue of Jesus dangling beneath a helicopter as it flies over the city of Rome.

It is tempting to work through each of Feenberg's cultural descriptions in detail. In this context, however, what is relevant is his conclusion that the dystopian approach to technological culture is ultimately unproductive, for it presumes, erroneously, the ability to stand outside the culture in the process of critiquing it. Proposals for action concocted by dystopian thinkers then tend to boil down either to total rejection or total renovation. Marcuse's visions, for instance, alternate between "the great refusal" and the idea of a radically other, pacified society. In contrast, Feenberg advocates an immanent critique. With Donna Haraway, Feenberg realizes that we need to work from the inside of the great technological beast; to put it another way, our task is not to find a way out of dystopia but rather to go forward from inside technological culture. Apocalyptic prophecies and dystopian images only cripple our ability to do this, for they obscure our ability to appreciate the mutability of technology.

Feenberg's final judgment regarding the dystopian beginnings of the critique of technology is ambivalent. He concludes that it amounted, in effect, to a first and necessary initial phase on the way to a full-fledged philosophical and social engagement with technological culture. Confronted with the phenomenon of technological culture—a unique event in the history of humanity, and which threatens to bring to an end the traditional rule of the intelligentsia and its power based on the symbolic interpretation of reality—critical thinking fled into a dystopian nightmare. It could free itself from this condition only gradually, and in this connection Feenberg remarks that the social constructivist approach to technology plays an important role in the process of liberation. Only when we think our way through dystopia and give this way practical form "will we find out what it really means to live and create in a technological society" (Feenberg 1995, 15).

Feenberg uses this sentence to conclude the introductory chapter of *Alternative Modernity*. Although the point is worth underscoring, it is also worth asking whether Feenberg hasn't been a little too quick in attributing the phenomenon of dystopia as a necessary phase in a sort of Hegelian process of consciousness-raising. Surely Hegel, and Feenberg, are right that it is useless to impose the present theoretical insights and moral standards on the past. Still, I think that Feenberg has failed to plumb the depths of the dystopian vision and of its negative effects deeply enough. I will not explore this point in depth here, but instead will restrict myself to a few brief critical remarks. It is striking that, in his discussion of dystopian thinking in *Alternative Modernity*, Feenberg fails to make any explicit criticism of his previous work, which despite critical discussions of Marcuse and Foucault could well be characterized almost in its entirety as dystopian. To be sure, in *Critical Theory of Technology* Feenberg imagines a theoretical way out of technological domination through the radically different technological socialist code. But as Feenberg himself admits, the practical prospects are poor—his references to Yugoslavia and China seemed hopelessly naive even in 1991—and he therefore winds up condemning us, just as did Marcuse and Foucault before him, to a dystopia we cannot escape. Another indication is his attacks on concrete technological and social reforms, which he dismisses as dangerously reformist inasmuch as they tend to be aligned with the dominant capitalistic logic. They are, to borrow a phrase from Marcuse, "coopted" into the system. Each protest strengthens the system; each concrete reform legitimates it. Although Feenberg does not say so, in such a perspective the theoretical opinion about the abolition of child labor can only be negative: Doesn't such an abolition in the end strengthen capitalism and make it more acceptable? Are we therefore to reject it? Fortunately, Feenberg is inconsistent enough not to embrace this kind of concrete judgment.

Thus the dystopian portrayal of technological culture has had, and still has, fatal consequences for those who attempt a truly critical engagement with technology and technological developments. The temptation is far too great for thinkers of the dystopian persuasion to dismiss reforms as "lubricating the system," and to reject and disengage rather than to participate in a critical dialogue concerning the content and direction of technology. Happily, reality usually intervenes, and theoretically inspired negativity often fades away, replaced soon enough by practically minded active engagement. This is also true of Feenberg's theoretical calls for a socialist alternative technology, as can be seen from the following example involving the analysis of a concrete technological development.

THE SPACE CREATED BY THE USER

Though it sounds bizarre nowadays, in the 1970s the computer played a starring role in dystopian doomsday scenarios. The computer seemed to represent the final stage toward the realization of the social panopticon of Bentham and Foucault, the technological agent that would make it possible to monitor and control each individual continuously from birth to death. During the 1970s and 1980s a large number of my students wrote papers on the social role of computers. They usually painted the march of this new technology as the ultimate threat to human freedom, as cementing the totalitarian character of modern society. At least until the mid-1980s, and especially in leftist circles, computers tended to be seen as dangerous and wicked instruments of the military, government, and multinational corporations. It hardly occurred to people that ordinary people would come to use computers, that computers would be useful to activists in coordinating and advancing their causes, and that computer use itself could be appropriated and transformed. But precisely this actually happened. Technological anarchists came to appropriate the hacker ethic, key values of which included free access to information via computers, decentralization of computer use, no-holds-barred discussion groups and chat rooms, and protection of the user's privacy. In the meantime, at least in part, many of these values have come to be realized thanks to the Internet.

In the fifth chapter of *Critical Theory of Technology* (1991), Feenberg addressed the fears then felt by the left about the large-scale use of computers. In it, and fully in line with his theory of the technological code described above, he perceived the computer as having emancipatory possibilities; the heading of the opening paragraph refers to "the ambivalence of the computer." Instead of seeing computers as inevitably leading to increased domination and to loss of jobs, Feenberg saw them as potentially in the service of unfolding new work possibilities and of self-organization. Feenberg referred expressly to

Dreyfus's ideas as providing philosophical support. But Feenberg's ideas remained exclusively on the theoretical level, and his descriptions of these other possibilities were couched in general and abstract language borrowed from socialist rhetoric, full of words like "participation" and "self-governance." Concrete examples of an alternative, socialist use of the computer were nowhere to be found.

In a strict sense this is also not the case in *Alternative Modernity*, in which two chapters are devoted to the computer. In one of these, Feenberg writes enthusiastically about the way users of the French telephone system appropriated Minitel for their own purposes. But they used these devices not in the lofty, politically correct alternative ways that Feenberg had in mind in 1991. Instead of using the new technology to move themselves in the direction of liberation, emancipation from work, and self-organization, they used it for gossiping and accessing pornography.

Feenberg's enthusiastic story about Minitel reveals that he has to a large extent forgotten his earlier theoretical and critical remonstrations about an "alternative," socialist technology. Still, his eager embrace of the way the French used the Minitel technology should not be negatively interpreted as backsliding. It is instead positive: Feenberg's example of Minitel is a perfect example of the transformation of the technological code—of how the development of a technology can be affected in important ways even after its introduction. The story of Minitel is worth a little elaboration in view of the way that it would later be recapitulated by the story of the international Internet.

In the beginning of the 1980s the French government distributed millions of free terminals to telephone users as part of a system called Minitel to provide access to central data and facilities and to make possible such things as telebanking. Neither the designers of the technology nor the politicians and theorists who paid attention to the project anticipated that it would have any use other than information exchange. That even philosophers lack a privileged vision of technological development is clear, Feenberg points out, from Jean-François Lyotard's famous book of 1979 entitled *The Postmodern Condition*, one of the earliest and most lucid studies of the impact of computerization. Lyotard foresaw only that the computer would provide the foundation for a future *information* society; it would be the means to achieve unrestricted access to a vast, centrally stored warehouse of information. Neither he nor any of Minitel's technical engineers anticipated that it would also be used as a means of *communication*.

And yet as early as 1982 diverse groups of hackers were using Minitel to communicate with each other instead of as an information-gathering device. French public officials were initially outraged by this "misuse" of the system, and there followed a public debate over how to define the new medium—was it for information or for communication—that was settled very quickly as throngs of French citizens used their terminals for social contacts. In effect, the system was altered by the users. In defiance of the clear and explicit intention of Minitel's designers, its users in effect were able to socially "redesign" the technology. The utopian vision of a perfectly functioning information society went unrealized, as did the dystopian vision of a panoptical society. Between utopia and dystopia, Minitel's users managed in an utterly unexpected way to create a contingent and, as Feenberg writes, *messy* new construction.

These kinds of developments, in which the users of a technological system wind up redesigning it, take place again and again in the interaction between high technology and contemporary society. The precipitous pace of the Internet's development has encouraged us all to forget that, until shortly beforehand, the computer was thought of *only* as a means of receiving or processing information. And this kind of basic developmental trajectory, though at a much slower pace, also characterized previous information technologies. For instance, as a related historical illustration of the underdetermination of the development of technology, the contingency of technological developments, and the impact of user groups, a brief reference to the early stages of the introduction of the telephone into The Netherlands will be instructive.[1] The telephone was initially conceived there as an instrument for conveying information of a business-like nature, and was frequently described as a sound or spoken version of the telegraph; the laws it was considered to fall under, for instance, were those pertaining to the telegraph. Its early applications, fully in line with this information-based conception, included using it to carry musical events and even church services (so-called church telephones). In the early stages of the telephone's development in The Netherlands, it didn't occur to anyone that it could or would become an important instrument of social exchange. But its users, and especially housewives, gradually adopted it as a conversational medium, though this "misuse" was initially strongly condemned. A telephone etiquette slowly developed according to which a telephone conversation should be short, polite, and functional. Some users (again, mainly housewives) strongly resisted this concept of the telephone as principally a means of information. In the United States, it took over fifty years before the role of the telephone as an important instrument of sociability was fully accepted. In The Netherlands, though, it was not until the 1960s that the telephone completed this transition, when its typical place was moved from the cold entryway of the house to the warm interior.

The subsequent history of the telephone would take us too far afield; what I've just mentioned should make sufficiently clear that the development of this medium was not determined solely by technological criteria but was under-determined, allowing housewives to put it to their own use as a social medium. This story has been repeated in its essentials in the case of many developments in information technology. Nowhere in these developments is there any trace of a technological necessity that inexorably drives them onward. And in the case of technologies still in their infancy, much room remains open for contingency. The development and use of answering machines, telephone services, mobile telephones, and picture phones does not seem to be headed inexorably in any prescribed direction. Users appear to have available many possibilities for reworking and redesigning even the most modern information and communication technologies. In this respect we live in an unpredictably messy world rather than in a normalized and regulated dystopia.

GO AS A METAPHOR FOR A TECHNOLOGICAL CULTURE

Feenberg makes clear that the story of Minitel and its changing fortunes could have unfolded in that particular way only in France. The story of Minitel required the specific background of French technological culture, which is characterized by a strongly centralized state that sees itself as responsible for guaranteeing the spread of specific technologies and the values associated with them. This is not, for instance, technological development the way Ellul sees it—an autonomous and natural development within a homogeneous, global technological culture in which each step is determined by the single, all-important goal of efficiency. Clearly there is not just "one best way" within a technological culture but many different ways associated with many different goals.

Although Feenberg only touches on this issue in passing in the Minitel story, he addresses it explicitly in the two chapters of *Alternative Modernity* devoted to Japanese technological culture. The first is devoted to the philosophy of the twentieth-century thinker Kitaro Nishida, who sought to formulate a distinctively Japanese definition of modernity. It is not necessary to elaborate this philosophy, which has many affiliations with that of Heidegger; what is salient here is that Nishida fully embraced Western science and technology, recognized that its introduction and evolution would bring about fundamental changes in a traditional culture such as Japan's, and yet believed that throughout the coming radical changes Japanese society could create its own form of modernity in which past cultural practices could still play a large role. His views thus strongly differed from such African and Asian political thinkers as N'krumah, Nyerere, and Nehru, who believed that, following decolonization, their lands could rapidly adopt Western technology without significantly altering their cultural traditions. Nishida realized the power of the moderniza-tion project that the introduction of Western technologies would entail, but he did not think it had to involve the kind of uniformity and homogeneous rationalization predicted by Max Weber, for instance, at the beginning of the twentieth century, who described an "iron cage" of technological rationality that would enslave alike all cultures worldwide. Instead, Nishida envisioned a technological culture that, while steadily growing and becoming more global, would simultaneously become pluralistic. Rather than slowly grinding tradi-tional cultures into oblivion, the new technological culture might be able to adapt in different ways to their different spirits and become, as it were, so many different experiments in modernity.

In the following chapter, Feenberg engages in a deep and thoroughgoing discussion of the game of Go in an extraordinary appeal to those who share the dystopian vision of the uniformity of technologically tinged criteria and values that unfold independently of a cultural context and even come to deter-mine it. Feenberg's analysis is based on the novel *The Master of Go* by the Japa-nese Nobel laureate Yasunari Kawabata. In outlining Feenberg's use of this ex-ample to illuminate theoretically the possibility of an alternative modernity, I shall borrow from both Feenberg's analysis and Kawabata's novel.

At first sight *The Master of Go* seems merely a literary treatment of an account of a Go competition in 1938 that Kawabata covered as a young jour-nalist. The competitors were an old master named Shusai and his young chal-lenger, Otaké. The match was sponsored by the newspaper for which Kawa-bata worked, and because of an interruption when Shusai fell ill it lasted many months. Both the novel and Feenberg's chapter discuss the rules of the game, as well as a number of events that played a central role in the match, which there is no space to recapitulate here. Suffice it to say that Go surpasses chess in terms of complexity and the number of possible combinations. Feenberg calculates that the number of possible moves is the factorial of 361, or more than the total number of atoms in the Milky Way.

The point of departure for both Feenberg's discussion and Kawabata's novel is that the game of Go can be understood as an autonomous region with its own formal logic. The ambiguity that belongs to ordinary life and language is ruled out by its rules and quantitative aspects. And inside the game both players are alike, abstracted from all cultural and symbolic differences. Feen-berg rightly claims that this strongly resembles characteristics of modern scien-tific and technological rationality. In many respects the game of Go resembles the modern rational institutions of the sort that Weber described, which are also characterized by rules, simple and unambiguous criteria, precisely defined spheres of application, a predetermined content, and the equality of all parties involved. In many theories of modernity this neutrality and formal rationality of modern institutions is contrasted favorably with the irrationality and inequality characteristic of traditional cultures.

On one layer of meaning, *The Master of Go* is a critique of this simple presentation. For while it is true that traditional Japanese society appears to have yielded in certain domains to formal rationality—a form of rationality that is not at all a prerogative of the West—Kawabata is attuned to, and manages to bring to light, the way that formal rationality can take on completely different substance in different historical, cultural, and social contexts. Formal rationality is not at all as context-independent as it might seem at first glance, and as Weber's analyses suggest. This becomes especially clear in the next layer of meaning in Kawabata's novel.

The *Master of Go* is often interpreted as a somewhat nostalgic metaphor for the downfall of a rich traditional culture as it inevitably gives way to the steamroller of cold modern rationality. Because of the huge—partly financial—stakes involved in the competition, each action in and surrounding the match is formally contracted, which appears tantamount to yielding to the formal Western way of doing business. This greatly irritates the old master:

> It may be said that the Master was plagued in his last match by modern rationalism, to which fussy rules were everything, from which all the grace and elegance of Go as art had disappeared, which quite dispensed with respect for elders and attached no importance to mutual respect as human beings. From the way of Go, the beauty of Japan and the Orient had fled. Everything had become science and regulation. (Kawabata 1972, 52)

To those who might read these sentences as indicating a confrontation between the rationalistic West and the "inscrutable" East, let me recall the rationality of the game of Go itself, which can assume different forms. Kawabata avoids glorifying the tradition over against modernity in myriad ways, referring to the loneliness of the Master and to the unhappy life he had led. He is anything but idealized; Kawabata even mentions a possible episode of cheating—of rule-bending—by Shusai that was instrumental in earning him his high reputation. In these and other ways Kawabata manages convincingly to problematize the clichéd picture of a confrontation between tradition and modernity.

In Japanese culture the game of Go has developed together with and related to Zen Buddhism into a "way of life." Despite or perhaps even because of its formal rules, it was a way of attaining a state of mental emptiness via extreme concentration and discipline. At the same time, the struggle against one's opponent brought about a work of art on the Go board; Feenberg even speaks in this connection of an "aesthetic achievement of harmony and pattern" that is achieved in the conflict. The modern approach exemplified by the younger Otaké, whom Kawabata portrays not at all unsympathetically, is motivated entirely by winning or losing.

Or so the Master thinks, at least. At several points he expresses anger at this modern mentality. Otaké, for his part, thinks that he is doing no more, and no less, than adhering to the formal rules. The tension between the two antagonists reaches a dramatic climax in the culminating stage of the match, when all the spectators are closely following a decisive struggle unfolding in the center of the board. Otaké's next move is near the top of the board. The Master and the onlookers are stunned when they first see it. The move causes a wave of revulsion and indignation to flood over the spectators—it is "unworthy" of a Go player, they say—though the Master reveals nothing of his feelings. Shortly thereafter he speaks out: "The match is over. Mr Otaké ruined it with that sealed play. It was like smearing ink over the picture we had painted. The minute I saw it I felt like forfeiting the match. Like telling them it was the last straw." But the hostility felt by the Master disrupts his concentration; he makes a false move, and quickly loses.

Both the unfolding drama as well as Feenberg's account reveal that Otaké's surprising move, "black 121," can be interpreted in many different ways, historically, aesthetically, and socially—"just as a technological artifact or a scientific fact," Feenberg remarks. As for aesthetics, "The Master had put the match together as a work of art. . . . A masterpiece of a game can be ruined by insensitivity to the feelings of an adversary" (164). But younger players think Otaké's tactic is entirely legitimate, for he has managed to prevail without violating a single formal rule. And the narrator-journalist in the novel is ambivalent: he recognizes that the changed cultural environment can also change the game of Go, and that despite the wholly formalized nature of its rules it is clearly not context-free. He describes the conflict between the Master and Otaké as "the end of an age and the bridge to a new age." For he notes that, even though Otaké's tactics involve a certain loss of beauty, they also amount to the introduction of a "new vitality" into the game. When Otaké at one point wants to forfeit the game, frustrated by what he sees as the Master's impossible attitude, the narrator-journalist remarks that "to forfeit the match would be to interrupt the flow of history" (145). Otaké, he thinks, bears "the responsibility for an emerging era" (146). Neither the new era heralded by the surprising move "black 121," nor the old world on behalf of whose preservation the sickly Master risks his life, comes across as unambiguously good or evil, or even as unequivocally superior to the other. Nevertheless, the many dystopian interpreters of Kawabata's novel typically ignore all the evidence of ambivalence and see *The Master of Go* as simply an elegy for the rich traditional culture of Japan as it is slowly crushed by the overwhelming, leveling, and oppressive power of modern, formal rationalistic methods.

The extent to which formal rationality can help foster historical developments, aesthetic values, and social interests is also clear from two passages that Feenberg does not mention. In the first, the narrator remarks that Go is not only on the verge of a new and quite different future; even its present form is strikingly different from the way it was played in the past. While its rules remained unchanged after the game was introduced into Japan from China, its meaning was significantly developed in the new cultural context. Kawabata discusses this historical background at length, and it is not far-fetched to see a reference to Western technology in his conclusion: "It is clear that in Go the Japanese spirit has transcended the merely imported and derivative" (118).

A second suggestive passage describes a game of Go played between the narrator-journalist and an American during a train ride. The foreigner has taken lessons at a Go-club, his training in the game is sound, and he has reached a passably high ranking. Still, with his different cultural background he plays the game in an entirely unfamiliar way. Kawabata felt "no response, no resistance" in his play. Despite his clear ability, "The spirit of Go was missing. I thought it all very strange, and I was conscious of being confronted with utter foreignness" (116). Thus despite the formal rules and reasoning involved in the game, Go is something quite different in Japan than in the West. Social constructivists would say that there are at least four different versions of Go— the original Chinese, the traditional Japanese, the modern Japanese, and the Western—just as there are as many different artifacts as there are groups that use a particular artifact differently.

The lesson Feenberg takes away from this discussion of the game of Go is the possibility of different alternative versions of modernity. A short recapitulation of his vision must suffice.

If we are seeking an alternative modernity that differs from the existing Western version, it has to have more content than the ethnic and cultural differences in, say, cuisine, juridical styles, or folk customs. Surely many of these and other practices are important to maintain, contra the prophets of modernization on the one hand or of dystopia on the other who see all such practices as soon to be leveled off. But Feenberg means to dig much deeper in his quest for the fundaments of a possible alternative modernity. In this connection he invokes the genealogical method of Nietzsche, who looked back at the words of Socrates and Jesus to uncover deep and hidden patterns in Western culture. In the same way, Feenberg says, we might seek to discover in traditional Japanese culture a number of assumptions and values that can be used to help fashion modern Japanese society into a quite distinctive technological culture. The persistence of such assumptions and values would not be anachronistic, but rather modern phenomenal forms of a genuinely historical culture. Feenberg mentions an example: "the culture of place." Each Japanese person—whether man or woman, master or servant—is assigned their own proper place in society. There is for instance "a clear enough distinction between the way in which men and women speak." Some grammar books "actually offer dialogues in both male and female versions." From the simpleminded perspective of the Western ideology of equality, of course, this appears to be a reification of inequality, the remnant of a traditional hierarchy—but for the Japanese themselves, Feenberg claims, this can be pictured as a sort of aestheticization of a dramatic struggle.

The tension at issue here cannot be neatly tidied up as a simple clash between modernity and tradition, between universalism and particularism. Feenberg describes the universalistic pretensions of modern Western culture as simply "false"; at best we may speak of a "practical universality." In practice modern society is inseparable from Western technological achievements, from electricity to antibiotics and especially in the worldwide networks of transportation and information. Within this kind of practical universality, however, "a multicultural politics of technology is possible" (Feenberg 1995, 232). With it, Feenberg sees the possibility of preparing a totally different future, one that departs significantly from the one projected by contemporary Western technological culture. "For that future, technology is not a particular value one must choose for or against, but a challenge to evolve and multiply worlds without end" (Feenberg 1995, 232).

A CONCRETE ALTERNATIVE

So runs the concluding sentence of *Alternative Modernity*, and it clearly testifies to a remarkable evolution in Feenberg's thinking. The marxist vision of *Critical Theory of Technology*, involving the conflict between two worlds—an evil capitalist old world and a good socialist future one—has disappeared. In *Alternative Modernity* its place has been taken by a more sophisticated "many worlds" theory in which technologies can help open up cultural, social, and technological possibilities in new pluralistic worlds. This striking evolution in Feenberg's philosophy was the result of a creative connection between critical theory and the social constructivist approach to technology.

In Feenberg's latest book, *Questioning Technology*, this connection is both broadened and deepened. The broadening consists of the fact that alongside his traditional social themes—Minitel, the Internet, child labor laws, patient participation in the testing of AIDS drugs—Feenberg now examines environmentalism and the role of technology. The deepening consists of the fact that he develops a fully elaborated social-philosophical theory of technology with the same thoroughness of the discussion of the game of Go in *Alternative Modernity*. "The time has come for an anti-essentialist philosophy of technology," promises the introduction to *Questioning Technology*. "We have had enough of timeless generalizations about technological imperatives, instrumental rationality, efficiency, enframing and similar abstract categories. This book offers a sketch of a concrete alternative" (1999, 1).

Questioning Technology has inspired some discussion among critics regarding the extent to which Feenberg has simply continued or truly revised his social-philosophical theory. The book does indeed creatively synthesize all of his previous work, and I shall provide a schema of its general outlines. Then I shall provide some concretization of this schema by reference to his treatment of environmentalism, by way of discussing to what extent the book fulfills its ambitious promise.

Feenberg characterizes his new general approach as an "instrumentalization theory," whose aim is to describe how a technical artifact becomes a usable instrument in society. The theory examines two levels of this process. The first level, or "primary instrumentalization," Feenberg sees as the area in which the classical, strongly essentialistically oriented philosophy of technology did its research. The work of Habermas figures on this level, as does that of Heidegger, whom Feenberg considers the most important thinker on technology of the twentieth century. On the second level, or "secondary instrumentalization," Feenberg places all the knowledge that the empirically oriented study of technology has developed in the previous decades. Once again we run across Latour as a great influence on Feenberg's thinking.

The first level of instrumentalization corresponds to the perspective of the classical philosophy of technology on modern technology, but also to the common sense conception of technology and the conception of technical experts themselves. This level concerns what Feenberg calls the "functional constitution of technical objects and subjects," and addresses the meaning of modern technology apart from all the social meanings that it might receive. That is not to say, Feenberg emphasizes, that social aspects are entirely absent from this level. The sharpness of a knife is obviously a technical property, but it is from the outset called for by and fulfills social interests involving the cutting or slicing of bread or vegetables, etc. This first level of instrumentalism, Feenberg says, consists of four moments of technical practice that "encompass the forms of objectification and subjectivation associated with a functional world relation" (1999, 203).

1. Decontextualization

The modern technological approach abstracts the natural object from the context in which it is originally found. Borrowing an image from Heidegger, the tree, to be technologically useful to the carpenter, must be cut down, removed from the woods, stripped of bark and branches, and turned into lumber, etc. In like fashion, the modern technological approach "de-worlds" the whole of nature, turning it into quantitative, calculable material for human projects.

2. Reductionism

This moment involves a kind of reduction. The decontextualized things are reduced to those aspects that can be enrolled in a technical network and that are essential to the technical program. Feenberg invokes a famous distinction of Locke's when he speaks of these technologically interesting aspects of things as "primary qualities." An object's "secondary qualities," which are of aesthetic, ethical, and cultural value, are unimportant to the technician who measures and calculates.

3. Autonomization

The subject of technical action, Feenberg says, "autonomizes" itself; it "isolates itself as much as possible from the effects of its actions on its objects." It rules over its objects and notices little or nothing of the world with which it is involved; it receives little or no feedback from the environment. "The hunter experiences a slight pressure on his shoulder as the bullet from his gun strikes the rabbit; the driver hears a faint rustling in the wind as he hurtles a ton of steel down the highway" (204).

4. Positioning

This moment amounts to an extension of Francis Bacon's dictum that "Nature to be commanded must be obeyed" to the nature of human beings. It is true that human beings do not obey laws in the same sense as do objects; still, if one "knows" human beings (or what one might call the "primary qualities" of the "human resource"), one can effectively manage and direct human behavior. Human beings, in short, are treated purely and simply as a labor force or as consumers to be manipulated.

The above four moments of primary instrumentalization characterize the basic technical relation. Classical philosophy of technology, Feenberg claims, is right to critique it but has gone astray by focusing exclusively on this aspect of technology and failing to move beyond it. For the classical philosophy of technology, this level of instrumentalization is the whole story of modern technology and its development. Modern technology, it claims, is out to reduce our entire world to a single, vast, calculable whole whose elements are one and all at the disposal of technological interventions. The rich culture of handicraft from the past is to be replaced by an abstract, cultureless approach to reality.

But the more recent and empirically directed studies of technology, Feenberg points out, have allowed us to see that primary instrumentalization is only part of the story of modern technology. Primary instrumentalization provides only the conditions for the existence of contemporary technology. In order for there to be an actual technological system or device, a second level of instrumentalization is necessary. "Technique must be integrated with the natural, technical, and social environments that support its functioning" (Feenberg 1999, 205). In the absence of such integration, there is no possibility of having actual, functioning technology. This second level, which can be distinguished from the first only analytically, also includes four moments:

1. Systematization

In primary instrumentalization technical objects are decontextualized, detached from their environments; but to work in a new environment these decontextualized objects need to be combined and coordinated with each other in a system and reembedded in that environment. Latour describes this process as an "enrolling" of objects in a network; Feenberg calls it systematization. A detached, free-floating technological object cannot function in the absence of socio-technological systems—and today, these systems have become globalized. An automobile, for instance, could not exist—it would be a meaningless object—in the absence of that vast and global technological system that includes oil refineries, gasoline pumps, highways, etc.

2. Mediation

In order to function in society, the technical object must be supplied with appropriate (aesthetic and ethical) secondary qualities. In the past a traditional craftsman usually did that himself, but nowadays the aesthetic features (i.e., design and packaging) of an object are at least partly the responsibility of the marketing and design division of the company that produces it. Still, Feenberg claims contra Heidegger, even modern technical products can have an "expressive design." In any case, a technical object could not function without these aesthetic and ethical "secondary qualities."

3. Vocation

The life of a traditional craftsman was characterized as a "calling," as a duty that gave form and meaning to his life. But for a modern maker or user of technology, life and work are no longer bound in the same way to each other, nor to tools and products. Empirical research, however, often finds that

	Functionalization	Realization
Objectification	decontextualization	systematization
	reduction	mediation
Subjectivation	autonomization	vocation
	positioning	initiative

Chart I (Feenberg 1999, 208)

workers' lives today are positively engaged by technological work environments. Feenberg points to this research, coupling it with a call for further changes in the workplace that would lead to fostering the participation of the worker. Feenberg's schema evidently has a normative dimension on this second level; and his description of secondary instrumentalization is at the same time an implicit call for an attention to and strengthening of its different moments.

4. Initiative

In this final moment Feenberg reaches back to those examples from his previous writings (such as the development of Minitel) that illustrate how users can appropriate technical devices and systems in a myriad of creative and innovative ways. On the primary level of instrumentalization the user is treated as passive; but on this level users can transform technologies and even use them in ways counter to what was intended. This moment, too, has a normative dimension, for Feenberg calls for technologies to be developed in ways that maximize the possibility for individual initiative on the part of the users.

• • •

Feenberg regards the environmental movement as "the single most important domain of democratic intervention into technology" (Feenberg 1999, 93). Often, though, green attempts to influence and change technology are held captive by the so-called tradeoff model. This model, which belongs to the level of primary instrumentalization, insists that any environmentally oriented reform of technology be at the same time economically feasible. Then opponents of the greening of technology ask, "But how can technological reform be reconciled with prosperity when it places a variety of new limits on the growth of the economy?" (94).

Feenberg argues convincingly that this is a false dilemma that arises out of a defective conception of technology. Only when technology is reduced to the functional aspects of primary instrumentalization does this dilemma arise. Once one perceives that technology always already incorporates social values thanks to its embedding in socio-technical systems—in Feenberg's terms, once one incorporates the level of secondary instrumentalization into the analysis—the dreaded dilemma melts away.

The child labor case, again, shows how apparent dilemmas can arise in processes of socio-cultural change, and how these can be resolved by new social values that eventually become obvious and taken for granted when integrated with the technical code. New values such as the prohibition of child labor can be opposed by social groups in the name of efficiency. Such groups often argue that these are surely noble values, but we have to recognize that their realization comes at a price and may indeed cost us our prosperity. Meanwhile, environmentalists, who frequently borrow their conception from classical philosophy of technology and whose heroes may include Heidegger and Ellul, sometimes go along with this reasoning. The limitations of this approach for practical politics are obvious. Feenberg's new and broader theory thus opens new possibilities for environmentalism in the struggle for the greening of technology against sheer economic efficiency by replacing the tradeoff model with a model of politics able to form new technical codes. The practical relevance of Feenberg's theoretical distinction between the two levels of instrumentalization is that it suggests the possibility of a future in which, according to the apt last line of his book, "technology is not a fate one must chose for or against, but a challenge to political and social creativity" (1999, 225).

NOTE

1. Much of the following is drawn from Bergman, Frissen, and Slaa (1995), published in a study by the Rathenau Institute that discusses the history of government interventions in communications technologies.

BIBLIOGRAPHY

Bergman, S., V. A. J. Frissen, and P. Slaa. 1995. "Gebruik en Betekenis van de Telefoon in het Leven van Alledag (The Use and Meaning of the Telephone in Everyday Life)." In *Toeval of noodzaak* (Contingency and Necessity), edited by Rathenau Institute, 277–328. The Hague: Rathenau Institute.

Borgmann, Albert. 1984. *Technology and the Character of Contemporary Life: A Philosophical Inquiry.* Chicago: University of Chicago Press.

Feenberg, Andrew. 1981. *Lukacs, Marx, and the Sources of Critical Theory.* New York: Oxford University Press.

———. 1991. *Critical Theory of Technology.* New York: Oxford University Press.

———. 1995. *Alternative Modernity.* Berkeley: University of California Press.

———. 1999. *Questioning Technology.* New York: Routledge.

Foucault, Michel. 1977. *Discipline and Punish*. Translated by Alan Sheridan. New York: Pantheon.

Habermas, Jürgen. 1968. *Technik und Wissenschaft als Ideologie*. Frankfurt am Main: Suhrkamp. Translated as "Technology and Science as Ideology" and published in *Toward a Rational Society*, edited by Jeremy Shapiro (Boston: Beacon Press, 1970).

Kawabata, Yasunari. 1972. *The Master of Go*. Translated by Edward G. Seidensticker. New York: Vintage.

Latour, Bruno. 1987. *Science in Action*. Cambridge: Harvard University Press

———. 1992. "Where Are the Missing Masses: The Sociology of a Few Mundane Artifacts." In *Shaping Technology/Building Society*, edited by Wiebe Bijker and John Law. Cambridge: MIT Press.

MacIntyre, Alasdair. 1981. *After Virtue*. Notre Dame, Ind.: University of Notre Dame Press.

Marcuse, Herbert. 1964. *One-Dimensional Man*. London: Routledge & Kegan Paul.

Rathenau Institute, ed. 1995. *Toeval of noodzaak* (Contingency or Necessity). The Hague: Rathenau Institute.

4. Donna Haraway: Cyborgs for Earthly Survival?

René Munnik

In 1985, Donna Jeanne Haraway, a professor of women's studies at the University of California at Santa Cruz, published an article entitled "Manifesto for Cyborgs: Science, Technology, and Socialist Feminism in the 1980s." Though a difficult and complex text, it brought Haraway instant renown, and ever since its publication her name has cropped up with greater and greater frequency in discussions about the role of science and technology in contemporary culture, especially among feminist circles.

In her article, Haraway invited her readers to consider reality through the eyes of a *cyborg*. A cyborg is a bionic being, partly human and partly robot—a being in which the border between nature and (technological) culture is blurred in a body that mingles flesh and titanium. Cyborgs figure in many popular science fiction novels and films, such as those in the *Robocop, Terminator,* and *Star Trek* series. Cyborgs are constructions lacking a fixed, pre-existing, natural identity; they are made, not born. Cyborgs are partly fictitious—but they also partly reflect the predicament of contemporary human beings, whose existence is inextricably bound up in a technological framework, whose lifeworld has become dominated by technology, and whose body is a field of possible and real technological suggestions and interventions.

Haraway uses a wealth of empirical data in her texts; nevertheless, her philosophy of technology is chiefly speculative in nature. She uses the image of a cyborg as a *speculum;* as an instrument with which she can bring to light the hidden mechanisms and possibilities of our social and political reality that are dominated by technoscience. Cyborgs, Haraway contends, offer us a glimpse into things that can disrupt our current essentialist modes of thinking. In these essentialist modes of thinking, our ethics, politics, and conceptions of social reality and historical progress are founded in an anthropology in which the essence or nature of human beings is fixed. In these ways of thinking, our ethics, political order, and so forth should ideally correspond to this unique and original human nature. But such conceptions are out of the question for cyborgs. They have neither an original, pre-existing nature nor a stable identity; they are endlessly transformable constructions. The identity of a cyborg is not a primeval essence, but rather a temporary halt in an ongoing series of constructions. This, Haraway finds, is precisely what makes the cyborg perspective so suited for shedding light on identity as a construction.

Identity-as-construction works on a large scale when one speaks about the "human condition" and assumes this to take the form of a white Western male, thereby defining what it means to be a human being and implicitly excluding non-white, non-Western, non-male human beings from participating in humanity. But identity-as-construction also works on a small scale. The personal identity of each individual human being is not some "originary self" from which springs one's thinking and behavior, and from which one can be alienated—an alienation that is necessarily supposed to be inauthentic. Rather, identity is a temporary halt in an ongoing life history, which a person continually reconstructs in his or her biography—meaning all the stories that one tells about and to oneself. Identity and biography, evidently, belong to each other.

IDENTITY AND BIOGRAPHY

In a short, autobiographical essay, Haraway wrote:

> I find myself compelled by the way we repeatedly rehistorcize ourselves by telling a story; we relocate ourselves in the present historical moment by reconfiguring our identities relationally, understanding that identity is always a relational category and that there is no such thing as a subject who pre-exists the encounters that construct that subject.... Identity is an effect of those encounters. (Bhavnani 1994, 21)[1]

This is a particularly revealing passage. Not only does it sketch out the background and motivation for her approach to cyborgs; it also shows her applying that approach to her own identity. Haraway, it seems, is a cyborg herself—in fact, we all are, she would claim. We can gain an idea of what she means from her own biography.

Donna Jeanne Haraway grew up in Denver, Colorado, in an Irish Catholic environment. She earned a degree from Yale University, and in 1970 began full-time teaching in the General Science Department of the University of Hawaii in Honolulu, where she initially devoted herself to biology. But in

Hawaii she grew acquainted with a completely different world, which gradually made her aware of the political, racial, historical, and geographical situatedness not only of herself as an individual, but also of the biological science that she was teaching. In Hawaii's polyethnic society composed of Japanese-Americans, Hawaiians, Chinese-Americans, and Samoans, in which Americans of European ancestry were a minority, she first became conscious of racial, colonial, and post-colonial social relations. This realization was underscored by Hawaii's role, in the early 1970s at the end of the Vietnam War, as the center of the Pacific Strategic Command.

From her middle-class Denver background, in short, Hawaii came as a culture shock. Initially, however, this new perspective barely registered in her intellectual development. In 1971 she and a female colleague began giving courses on the biology and psychology of sexual differences, but, as she would later say, in an *unmarked fashion*. That is, she and her colleague weren't aware of the problematic of the racial differences between women, and taught women's studies from the unproblematized perspective of a white Western female. At the same time she began a professional evolution from biologist to historian of biology. This was another crucial point of departure. For historical research, especially when guided by a feminist political sensitivity, can bring to light the role of very significant human choices in the development of science, choices that, once ensconced as dogmas in the orthodox practice of science, rarely came up again for discussion.

Haraway's stay in Honolulu came to an abrupt end in 1974, when her husband, who was gay, lost a tenure case in which, she says, homophobia played a role. Haraway and her husband had already decided to terminate their marriage, which proved to be not a good forum for their intense friendship and sexual relationship. After this traumatic experience with homophobia, they left Hawaii, separated, and each found a job elsewhere. Haraway obtained a position in the History of Science Department at Johns Hopkins University in Baltimore. In contrast with the University of Hawaii, Johns Hopkins was a powerful and wealthy university, one that carried out important work in defense and nuclear research. Its medical faculty was outstanding, and its political science department played an influential role in the development of U.S. foreign policy. At Johns Hopkins, she was fully and freely encouraged to develop her interests in the history of science. At the same time, she was very conscious of the economic and social circumstances that had made this development possible: "Again the kind of money that was available in the United States, through the 1960s and 1970s just can't be underestimated in making people like me into intellectuals" (Bhavnani 1994, 21–22). For Haraway, the pretense of innocence was gone forever.

In this period she also became a member of the Women's Union, a biracial

(African-American and European-white) socialist-feminist organization. Thanks to it, the most important components of her later thinking began to come together: history of science, marxist philosophy of history, feminism, and the struggle against racism. "[I]t was a period when, as a historian of biology, a historian of science, I became much more aware of the history of Marxist discourse in the history of science, the development of feminism in science-related issues, the importance of the struggle against racism embedded in the history of science, the *very* important radical science history against the electronic battlefield, chemical and biological warfare, biological racism and sexism—those kinds of issues became the core of my intellectual work, my writing" (Bhavnani 1994, 22).

This new departure did not show up immediately in her writing. In 1976 she published *Crystals, Fabrics, and Fields,* in which there is scant trace of her new feminist political involvements. It is a solid but unsurprising analysis of twentieth-century biology, analyzing the work of three key figures: Ross G. Harrison, Joseph Needham, and Paul Weiss. In the book, Haraway sought to discover whether in twentieth-century biology the older mechanistic and vitalistic models had been replaced by organistic metaphors and models, and whether this change could be thought of as a paradigm shift in the sense articulated by Thomas Kuhn (Kuhn 1970). This book contained no indications of the political agenda that Haraway by then had set for herself. Marxism entered the book only in passing, in connection with Needham's Christian socialism, which adopted dialectical materialism as an alternative to earlier, more static ways of thinking, both mechanistic and vitalistic.

However, the book does provide a glimpse into the ways in which Haraway practiced history of science. Using historical analyses, she shows how the concepts, methods, and models that compose the orthodoxy of an established "normal" science—in this case, biology—are the contingent outcome of a historical process that constitutes the character and identity of that science. Internal scientific motives and arguments play a role in this process of construction—but also political interests and religious opinions, as well as unarticulated conceptions about the relations between races, classes, the sexes, nationalities, nature and culture, and so forth. In the book Haraway also points to the significance of metaphors in this constructive process. The identity of a science is the temporary outcome of a network of stories that it tells about itself and about its objects. In this sense the identity of a science resembles that of a person.

In 1978–79 Haraway published the first articles in which she is clearly politically engaged: "Animal Sociology and a Natural Economy of the Body Politic, Part I, a Political Physiology of Dominance" (1978a), "Animal Sociology and a Natural Economy of the Body Politic, Part II, the Past Is the Contested Zone: Human Nature and Theories of Production and Reproduction in Primate Behavior Studies" (1978b), and "The Biological Enterprise: Sex, Mind, and Profit from Human Engineering to Sociobiology" (1979). These articles subsequently formed the basis of the first part of her book *Simians, Cyborgs, and Women* (1991). Here her history of science is clearly politically inspired from the outset, and directed toward a particular field of research: the history of primatology, and specifically the ethology of humanoid apes.

"It was in that period that I started working on *Primate Visions*," Haraway once remarked (Haraway 1989).

> I began to think of primates as "figures," as these germinal entities into which many people's imaginations are condensed. I saw primates as these creatures on the boundary between what counted as nature and culture, onto which a great deal of racial discourse was projected in the United States in the 1960s and after. I began to see how primates were part of popular cultures, movies, technical field studies, part of social psychology, part of evolutionary biology, and part of zoo management. These creatures existed at the boundaries of many constituencies that figured and carried the meanings of many kinds of stories in their bodies. The western animal/human boundary story was, from the start, a heavily racialized story. Think of the entry of the white women to Africa, women such as Jane Goodall or Dian Fossey,[2] who became the surrogate for man, and who went to make contact with the animals, across the chasm between nature and culture. They were figured as *alone* in nature. And this was happening in the early 1960s, when the very areas of the world in which these animals lived were gaining national independence as the culmination of decolonizing struggles. So fifteen African nations achieved national independence and UN membership the year Jane Goodall goes to nature to be alone with the chimpanzee? . . . She goes at just the moment of repossession of territory on the part of African nationalists, who, of course, developed very different stories about the plants and the animals inside their own national boundaries and very different stories about the peoples and about ethnicities. (Bhavnani 1994, 22–23)

Haraway shows how the relation between nature and culture is defined and constructed in scientific studies of humanoid apes—whose existence straddles the boundary between human beings and other animals—and how concepts of race, class, and sex play important roles. This occurs, for instance, when field research is depicted as an idyllic meeting of white male humans with nature, transpiring in an empty landscape where the native inhabitants appear as stage props; or when Western museums display animals in "authentic" scenes in dioramas; or in attempts such as that of Professor Calmette of the Pasteur Institute, who in 1924 tried to civilize chimpanzees through an educational program in which native women were assigned the roles of waitresses and nannies.

In each case the primates functioned as "figures"; as blanks onto or around which were projected various meanings—of human beings, of animals, of women, of "man," of black, and of white.

In 1980, Haraway left Johns Hopkins and took up a position as a feminist theorist on the History of Consciousness Board at the University of California at Santa Cruz. The university's History of Consciousness program is interdisciplinary and even, Haraway says, "antidisciplinary"; that is, it disrupts well-defined demarcations between different scientific and philosophical disciplines and literary genres by making them interfere, thereby contaminating them with each other's language games, and creating an unfamiliar but revealing discourse: a heteroglossia. Haraway had already done this, for instance, in her research on primatology, which combined scientific data, museum representations, and also images drawn from popular cinematography. In Santa Cruz she continued her research on primates, but simultaneously began to develop a second figure: the cyborg. The two figures are related, for both are border-crossing figures. But while the primate is situated at the border between humans and animals, the cyborg is situated at the border between humans and technology. From 1985 up until the present the cyborg has been a dominant figure in Haraway's writing.

As should be clear even from this short biographical sketch, Haraway considers her political identity, and her oppositional thinking and activity, to have arisen from a number of very concrete, diverse, and situated encounters and adventures. The cyborg is, as it were, the central figure/representation in her thinking and activity, but it also embodies the situatedness of every identity. The rest of this chapter will therefore be devoted to the figure of the cyborg in her writings, and its role in her reflections on the meaning of technology.

THE CYBORG: CORE CONCEPT OF A POSTMODERN ANTHROPOLOGY

The seminal article, "Manifesto for Cyborgs: Science, Technology, and Socialist Feminism in the 1980s," which appeared in 1985, had a short prehistory. It grew out of earlier studies she had done for her contributions to a conference in April 1983 at Barnard College: "New Machines, New Bodies, New Communities: Political Dilemmas of a Cyborg Feminist," and "The Scholar and the Feminist X: The Question of Technology." That October she finished an article that would become an early version of the cyborg manifesto, which appeared in 1984 in Germany in *Neunzehnhundertvierundachtzig* (Haraway 1984).[3] But there are striking differences between this early version and the later cyborg manifesto. While the early version is addressed almost exclusively to genetic engineering, and the concept of the cyborg is introduced almost at the very end, in the final version in *Socialist Review* the concept plays a much greater role, together with other technologies, especially information technology. This final version of 1985 also appeared, under a slightly different title, in *Simians, Cyborgs, and Women*,[4] followed by other articles that take up the theme of cyborgs.

But what precisely is a cyborg, and what does Haraway think she is bringing to light with this concept? A straightforward answer to this question is impossible because cyborgs themselves lack a single distinct meaning or identity, and according to Haraway are not *meant* to have any. Cyborgs, she insists, have a "broken identity." This idea can be clarified by discussing three different facets of the cyborg concept.

1. Origin and Extension of the Cyborg Concept

The term *cyborg* was coined, not by Donna Haraway, but rather by Manfred Clynes, a space travel researcher, and Nathan Kline, a psychiatrist, in 1960 (Clynes 1960; Haraway 1996, 51–52). Clynes and Kline proposed that the physiological aspects of space travel would require the development of "self-regulating man-machine systems," and named such systems "cyborgs," a bastardization of "cybernetics" and "organism."[5] They actually developed an example of such a system by implanting into a laboratory rat an osmotic pump that continually injected chemicals into the animal's bloodstream—thus, a living creature whose digestive system included a technological accessory. In this meaning of the term, human beings have already become cyborgs with the advent of such things as implanted insulin pumps and pacemakers, artificial heart valves, hip joints and eye lenses, and so forth, not to mention the temporary taking over of an organ's functioning by a dialysis or heart-lung machine. Technology is already grafted on to us humans; it is already embodied in the literal sense of being under our skin.

But the term *cyborg* can mean more than these routine facts, for it is also a figure in technological expectations. Consider the article "Will Robots Inherit the Earth?" by Marvin Minsky, a professor of artificial intelligence at MIT. The answer is already contained in its subtitle: "Yes, as we engineer replacement bodies and brains using nanotechnology. We will then live longer, possess greater wisdom and enjoy capabilities as yet unimagined" (Minsky 1994). Minsky goes on to paint a compelling picture of a technological future in which science and technology overcome the limitations of our naturally endowed body through gene replacement, organ replacement, and the extension of human brain power with the help of information processing accessories, connected via electrodes with the corpus callosum, the largest databus in the brain. In this technological future, natural selection will be superceded by a consciously directed and unnatural selection process, producing cyborgs in order to realize ultimately unimaginable (human?) abilities.

Minsky then adds: "Needless to say, in doing so we will be making ourselves into machines. Does this mean that machines will replace us? I do not feel that it makes much sense to think in terms of 'us' and 'them.' I much prefer the attitude of Hans P. Moravec of Carnegie Mellon University, who suggests that we think of these future intelligent machines as our own 'mind-children.'" Cyborgs, Minsky says, are "mind-children," creative products that spring from the "spirit" or "soul" of modern technology. In this meaning of the term, cyborgs are (still) fictitious, the stuff of science fiction. But fictions can be compelling, and able to mobilize large amounts of capital and effort toward their realization. Cyborgs can be very real; that is, as yet-unrealized goals of technological development they can be supremely effective. Haraway has this meaning in mind when she speaks of cyborgs as "the awful apocalyptic *telos* of the 'West's' escalating dominations of abstract individuation, an ultimate self untied at last from all dependency, a man in space" (Haraway 1991, 150–51).

But besides the factical and fictitious meanings of the term *cyborg*, Haraway wants to invoke a still further and more fundamental meaning: a being whose context, lifeworld, social relations, and self-interpretation are thoroughly permeated by modern technology. In this last meaning human beings are one and all cyborgs. "By the late twentieth century, our time, a mythic time, we are all chimeras, theorized and fabricated hybrids of machine and organism; in short, we are cyborgs. The cyborg is our ontology; it gives us our politics" (Haraway 1991, 150). Although Haraway incorporates all three meanings of cyborg in her work, she emphasizes this third one, which marks a fundamental turning point in philosophical anthropology. Philosophical anthropology is generally conceived as anthropo-ontology; it reflects the ways of being of humans. But at the end of the twentieth century these ways of being are inextricably involved with technology: anthropo-ontology is cyborg-ontology. But what does *that* mean?

2. Cyborgs: Chimeras and Monsters

In the remark cited above, Haraway describes our time as "mythic." In myths we frequently encounter chimeras, hybrid beings that are half-human, half-animal, or composed of different kinds of animals: centaurs, harpies, Pegasus, the Minotaur, the Sphinx, satyrs, gorgons, Nereus and the nereids. These beings embody a blurring of the boundary between animals and humans. In their bodies different identities are brought together, but not as a metaphysical union or dialectical *Aufhebung* of the differences in the form of a higher unity. Rather, they are mixtures, bastardizations, corrupt collections of heterogeneous elements.[6]

Haraway proposes that we should imagine cyborgs similarly, as beings that blur boundaries, bastardizations of humans and technology. This was already the view of Clynes, Kline, and Minsky. But Haraway has much more to say. The discussions of Clynes, Kline, and Minsky reveal the cyborg as blurring the boundary between fact and fiction. This much is not new: cyborgs figure as easily in the most trashy SF-comics as in the most cutting-edge scientific research, where they are not only talked about but realized. But for Haraway there are three further fundamental *boundary breakdowns* to be identified, which find their origin in science and technology and which lie at the foundation of the ontology of cyborgs. The first is the erosion of the difference between human and animal brought about by modern biology. The second is the undermining of the difference between the animal-human organism and the machine. Pre-cybernetic machines did not yet exemplify the principle of self-movement or self-regulation, but contemporary machines are autonomous in a way that makes us no longer sure of the difference. The third boundary breakdown is that between physical and nonphysical. By this Haraway means not so much the bridging of the gap between mind and matter but rather the fact that contemporary technology, although obviously a material reality in a strict sense, by miniaturization and informatization tends to make itself more and more invisible and intangible ("immaterial"), thereby gaining a power that becomes increasingly uncontrollable.

Science and technology lie at the basis of this boundary blurring, and this is as true from the point of view of theory—we no longer know where to draw the exact boundary between human and animal, nature and artifact, material and immaterial reality—as from the point of view of practice—we are now in the position of being able to realize endless fusions or bastardizations between such heterogenous components. One result, for instance, is the OncoMouse™ (Haraway 1996, 78–80): a strain of mice, for which the firm DuPont de Nemours requested and received a patent, whose genetic material, invisibly, includes information from a human breast-cancer gene. A living being as an *invention;* an animal that carries a human carcinogen! The OncoMouse™ is a contemporary chimera, one that can be purchased for a mere hundred dollars each.

Like mythical chimeras, cyborgs have something monstrous about them. They embody something un- or anti-natural. They evoke revulsion in us because they muddle or defile the natural order of things. That is, they evoke revulsion from those for whom something like a natural order is still a fundamental value. But they are also monstrous in the sense that they de-monstrate something. As human constructions, they reflect the self-image and situation of contemporary human beings, whose lifeworld is shaped by the integrated circuit[7] of science and technology. In this self-image, a human being no longer understands him or herself as a particular instance of a unique and original human essence, but as a historical construction that is infinitely reconstructible and transformable. In cyborg ontology, a being—whether a human, man, woman, animal, nature, culture—is conceived as a constructed and constructible being. That means it is characterized by a radical contingency and situatedness.

3. Cyborgs: Contingency and Situatedness

The cyborg is as much a wish fulfillment as the partially realized product of the Western project of domination over nature. But as a produced and constructed Being—that is, one that has *neither* been created *nor* born—the cyborg has no origin myth, no Genesis. For such a being "the return to the origin" is a hollow phrase; it can never dream of a lost paradise. In Haraway's words "The cyborg incarnation is outside salvation history. . . . The cyborg would not recognize the Garden of Eden" (Haraway 1991, 150–51).

This thought requires some elaboration. The Western conception of history and progress is governed to a great extent by a particular figure of thought called "archaeology-teleology." History is conceived as a story with a plot. That story involves an original—pre-historical—unity or *arche,* and a fall away from this original unity. This fall begets history as such, a history that thenceforth is characterized by the dream of a lost unity. The completion of history would consist of a complete recovery, a repeal of alienation, and a return to the original unity. This completion is the goal (*telos*) of history; with it, the circle is complete. It is not difficult to recognize the religious sources of this structure: creation, fall, salvation, history, redemption in the New Jerusalem. Haraway often hints at those themes in her texts.

There are numerous versions of this plot line. In the optimistic versions, history culminates in an enriched return to the origin; this is history with the drama of a universal epic. In the tragic versions, the fall (the alienation from the original unity) is necessary in order to give birth to human subjects, who are then tormented by the dream of a lost innocence that cannot, however, be regained without regression or loss of self. Archaeology-teleology is found in each human project in which the final goal is determined by the origin. For instance, it is in play in the sheer obviousness with which human beings conceive their self-realization as a recovery of their own authentic self. Even classical conceptions of knowledge are characterized by such a movement, such as Aristotle's dictum that "What is first from the point of view of nature is the last from the point of view of knowledge." This means that the final goal of knowledge consists in the revelation of the beginning, the principle, the fundament of its object. For Haraway, the marxist conception of self-realization through work and the psychoanalytical conception of subject-formation through resolution of the Oedipal complex are nothing but secularized versions of these origin myths, in which "having a goal" is linked with "having an origin."

Because a cyborg can conceive itself only as a construction, it can never conceive itself as a moment in a history that involves recovery of an original unity or innocence that existed before a Fall. A cyborg never has such respectful or nostalgic sentiments. A cyborg is radically excluded from Paradise, and is untroubled by that fact. It is a being of everlasting dispersion; it has forever lost its innocence. For Haraway, this is the subversive character of the cyborg. But with the loss of its original innocence, it has also abolished an original guideline that serves to determine the Enemy, which can be anything or anyone representing the situation of alienation. Neither is the cyborg susceptible to utopias involving recovery of an original unity, utopias that, according to Haraway, will inevitably change into dystopias involving a deadly unity.[8] Cyborg existence consists in the art of surviving in the diaspora, a form of existence characterized by radical finitude and contingency (Haraway 1991, 170). For Haraway, therein lies the cyborg's political force. I shall return to this point in a moment.

Origin stories are not only about a paradise to be regained; they are also unity stories, and in a double sense. The first is that these stories claim universal validity. But all this talk of origin myths and universality founders for cyborgs. Cyborgs are radically factical, situated, and partial, without needing to have their own particular identity taken up into a "higher unity." Cyborgs are adverse to grand narratives. But origin stories—and this is the second sense in which they are also unity stories—are also about wholeness, integrity, and purity. Origin stories, for instance, typically bear upon the unity of the subject, the wholeness and authenticity of its identity, the purity and authenticity of types of discourse, an original order of things, and the generic principles and proper rules that hold within each domain and especially about the prohibitions and taboos of transgressing the boundaries of the domains. But cyborgs are transgressive of boundaries; they are corrupt beings; chimerical monsters in the sense outlined above. From the perspective of the ideal of purity and authenticity, these beings are perverse. Their identity is a broken, bastardized identity, a fusion of incompatibles. They are not unambiguously identifiable as "man" or "woman," "nature" or "culture," "human," "animal," or "machine," and they even do not shrink back from bestiality as a new form of marriage exchange (Haraway 1991, 152).

This point can also be expressed with reference to language. Cyborgs have no use for an all-encompassing, universal language game.[9] But neither do they accept a multiplicity of heterogenous language games, each of which underlies its own cleanly marked-off domain with its own bundle of rules or form of rationality.

Rather, cyborgs play the game of intertwining language games. From the standpoint of the pure and proper use of language, the language of cyborgs is corrupting, a heteroglossia. That means they have a talent for constructing "illegitimate" relations between language games with a possible result being the creation of powerful and incisive stories that are not as simple or easy to identify as those of science or fiction, social analysis or historical description, biography or poetry. Haraway's manifesto is itself this kind of story, which transgresses the borders between myth, science, technology, body, and language.

CYBORGS FOR EARTHLY SURVIVAL: CYBORGS AND POLITICS

The cyborg as the avatar of boundary blurring—part fiction and part fact, part nature and part culture—forms the central figure in Haraway's representation of the postmodern human condition, as dominated by the integrated circuit of technoscience. Such a monstrous portrait of the human condition might tempt us to think that Haraway is merely adding another voice to the rising clamor of technophobic prophets of doom, who fear any display of technology and who are ultimately motivated by a nostalgic longing for a return to nature. Another temptation might be to go the opposite way, and rank her among those postmodernists for whom the lack of a "nature," of "true reality," or of a "universal language" leads to open flirtation with the aesthetics of meaninglessness.

Neither would do justice to what Haraway is up to. She sees the former temptation, technophobia, as implicitly guided by naturalistic essentialism, a position that she unambiguously attacks. And she sees the latter temptation as paralyzing political activity by undermining it from the start with cynicism. In an article for *Configurations*, she describes her project of deconstructing original nature with the aid of the cyborg figure as follows: "Queering what counts as nature is my categorical imperative. Queering specific normalized categories is not for the easy frisson of transgression, but for the hope for livable worlds" (Haraway 1994, 60). And the heading of a subsection in *Simians, Cyborgs, and Women* is entitled "Cyborgs for Earthly Survival." Apparently, not all utopias are antithetical to Haraway's thinking; only those that offer hope of a return to an innocent past can count on her unsparing critique.

Indeed, Haraway's politics are present in every single page of her post-1978 writings: the analysis of power workings and power relations that lead to the cyborg, and of the new kind of power workings in a cyborg world—and all this from a socialist, feminist, and anti-racist perspective. In this sense she has always remained faithful to the political program that took shape for her in the mid-1970s.

How, then, does the cyborg figure in Haraway's political analysis? Here, too, figure a number of related points, but first a preliminary remark: Haraway's cyborg thesis is not meant as a philosophical, historical, sociological, or political position, which one can accept or reject, take a stand against, or use to initiate a debate. Rather, her cyborg thesis is a description of an anthropological condition in which political issues are at stake. A cyborg, for her, is not a political theme or slogan, but rather a signpost that points to novel ways of political conduct.

1. The Failure of Naturalism

It should already be clear that the cyborg—the denaturalized being par excellence—is meant to drive a stake through the heart of naturalism. But to grasp the relevance of this attempt we first need to take into account the role of naturalism in the consciousness of its victims. Naturalism, as Haraway conceives it, represents as necessary what is really contingent; it defines as original what is actually the product of a historical play of forces. But for Haraway, definition itself is a form of the exercise of power. Haraway finds this at work, for instance, in the Western image of man, which naturalism takes to be universally valid and applicable to each individual person, the basis of the "essential" unity of all human beings. But the Western image of man is, according to Haraway, a historical construction, the product of a play of forces in which this particular image arose and gradually became adopted as normative for "human being."

This Western image of man/human beings is not, however, an explicit, philosophically rigorous determination of human nature, but rather a figure in the sense in which the primate and the cyborg are figures—something that is implicitly present in the orchestration of narratives, practices, and theories about reality, that becomes visible only when one explores what is not in the narrative, which motives behind the practices are not made explicit, which perspectives in the theories are not brought up for discussion, and what things in it are not taken into consideration. In short, the unity of this image of man is the product of many processes of control and exclusion. But this image of man is not only the product of control. It bears control in itself as an essential feature, one that is reproduced and intensified by science and technology—man as a Cartesian "lord and master of nature"—that realizes itself as such through science and technology. The unity of this image of man, considered by naturalism to be the ontological basis of the alleged universal "we" of all human beings, is produced by excluding "outsiders" as an indefinite group of "others." Haraway leaves no doubt about her political agenda when she identifies the exclusions brought about by transnational capitalism, colonialism, racism, and sexism.

Does this mean that Haraway is setting herself up as *the* advocate, representative, and activist for all these groups of outsiders, in whose name she is writing? The answer is a resounding no. For a "natural" common group feeling or language, on the basis of which she could pose herself as representing a group, is exactly what these outsiders lack. And it is precisely at this point that the political relevance of Haraway's cyborg thesis emerges. Central to it is what one could call the irony of history.

2. The Irony of History

The awareness that science and technology are characterized by domination and control, and that the culture in which they have emerged and flourished has a patriarchal character, is hardly Haraway's own discovery.

However, she points out that the feminist opposition against it initially took the form mainly of the search for the essence of being female as the ground of women's own common identity, on the basis of which they could operate in a politically effective manner. But this attempt necessarily failed. "There is nothing about being 'female' that naturally binds women" (Haraway 1991, 155). Feminists soon discovered that each attempt to anoint an original essence reproduced the same mechanisms of exclusion and control as belonged to what they were repudiating: it usually took the form of the white, Western, intellectual middle-class woman who presumed to be able to speak in the name of her black "sister," suggesting an affinity where there was none. The outcome was a painful fragmentation among feminists along practically every conceivable fault line of ethnicity, class, nationality, and so forth. This brought Haraway to the realization that "identity" is in fact a sort of conglomerate resulting from a fusion of identities. "For me—and for many who share a similar historical location in white, professional middle-class, female, radical, North American, mid-adult bodies—the sources of a crisis in political identity are legion" (155). But none of this undermines the fact that the urgency of a mutual affinity is recognized all the more, although this affinity does not rest on a naturalist foundation.

So far, in this reflection on outsiders, we have been concerned only with Haraway's point that they have been brushed aside and excluded in the creation and perpetuation of the Western image of man. The irony of history is that outsiders have not only been created as the "other" of man/human being, but that—and at the same time!—an outsider figure has been created by the scientific and technological avant-garde as its realization, as its *telos:* the cyborg. To put this another way, Haraway refers to an "essential relationship" or "kinship" between, on the one hand, those existing beings who have been excluded by the realization of contemporary science and technology (the non-male, non-white, etc.) and, on the other, a new kind of being that has been brought into existence in this very realization (the cyborg). The concept "essential relationship" is inappropriate, of course; the essential relationship consists precisely in the fact that these beings cannot have an essential relationship in the naturalistic sense. Still, Haraway is proposing something akin when she suggests that "women of color"—the figure indicating the broken identity of the outsider—can be conceived as a cyborg identity, a powerful subjectivity composed of fused outsider identities (Haraway 1991, 174).

Nevertheless, Haraway's meaning is clear: the death of naturalism, which is represented by and embodied in the chimerical character of the cyborg, rings in an entirely new situation: that of the integrated circuit of high-technological culture. This new situation is characterized by a different logic of domination than the naturalistic one of control-by-normalization, so that the situation of the "outsiders" radically changes because the absence of an original model to identify with is no longer a sign of political impotence but instead an essential feature of culture itself. Haraway sets herself the task of exploring the possibilities of this new situation.

3. Cyborg Politics

It would be a complete misunderstanding of Haraway to say that she is led to these views by a naive faith in cyborgs and the "New World Order," as if the latest gadgets popping out of high-tech laboratories constituted "sister outsiders" riding to the rescue, or represented a new type of revolutionary class. Haraway harbors few illusions here. Ultimately, she says, cyborgs are illegitimate offsprings of militarism and patriarchal capitalism; modern warfare is a veritable "cyborg orgy" (Haraway 1991, 150), which should be sufficient to identify where they come from. But, she adds (with a sideways glance at Mary Shelley's *Frankenstein*), "illegitimate offspring are often exceedingly unfaithful to their origins" (151). In place of lost innocence, she claims, new possibilities loom for denatured beings—and she stresses the importance of investigating these possibilities, both for their livability and for their horror.

These new possibilities are couched in cyborg ontology, which Haraway specifies further as the "art of surviving in diaspora," or again as "play," a deadly serious play of life and death that deserves to be conducted with the proper sense of responsibility. At this point Haraway's approach is a very prudent one, as can be observed repeatedly, beginning with the way she characterizes her own relation to constructivism. Her constructivist bent should be clear from the foregoing: outsiders are the constructed "others" of man/human being, and cyborgs are constructed and constructible beings. Ultimately she will recommend that women play along with the pleasurable game of boundary blurring and the construction of hybrid identities.[10] But this constructivism is qualified. In writing about the social construction of science—an approach with which she heartily agrees—she observes that this should not be taken to imply that every (scientific) proposition reduces to pure rhetoric or to a strategy of persuasion.

> So, I think my problem and "our" problem is how to have simultaneously an account of radical historical contingency for all knowledge claims and knowing subjects, a critical practice for recognizing our own "semiotic technologies" for making meanings, and a no-nonsense commitment to faithful accounts of a "real" world, one that can be partially shared and friendly to earth-wide projects of finite freedom, adequate material abundance, modest meaning in suffering, and limited happiness.

She adds: "We need the power of modern critical theories of how meanings and bodies get made, not in order to deny meaning and bodies, but in order to live in meanings and bodies that have a chance for a future" (Haraway 1991, 187).

Constructivism concerning (scientific) knowledge is cyborg epistemology par excellence. It implies recognition of the impossibility of a naive realist correspondence theory of truth involving "true objectivity" as an ultimate court of appeal. Critical cyborg epistemology is an indispensable tool in the struggle against the claims of domination legitimated on "objective knowledge." But this explicitly does not mean that each truth claim evaporates! It receives rather the hermeneutic normative significance of truthfulness instead of blunt objective truth.

The normative significance appears when Haraway says that the game of constructing broken and hybrid identities must be responsibly played (Haraway 1991, 150). First of all, it should be noted that, as in games of love and power, rules for this game cannot be given in advance. In other words, it is not normed by a canon anchored in an extra-historical "nature of things." But this expressly does not mean that the subject is delivered over to a free-swinging boundlessness. Rather it means that this responsibility ought to find its expression in the fallible and situated tact with which the process of construction is carried out.[11] It would therefore be a big mistake, she says, to think that the deconstruction of original identities means that the subject—as a moral subject—disappears from the stage. Haraway will have none of the poststructural-ist "death of the subject."

Still, there is something to be said for the idea that her cyborg thesis leads to a decentering of the subject. Haraway's ideas do not empty out the concept of the subject; rather, they broaden it and thereby detach it from its exclusively human character. This becomes abundantly clear, for instance, when she speaks of all the different kinds of actors that play roles in the process of construction, which include not only human actors but also technological artifacts. The ability of cyborgs to blur and confuse boundaries means for Haraway that the boundary between man/human being/subject and thing/object can no longer be sharply drawn, meaning in turn that the domain of "corrupt" subjectivity is extended far beyond human beings. This is clear in her remark that "OncoMouse™ is my sibling, and more properly, male or female, s/he is my sister" (Haraway 1996, 79). This remark is a form of heteroglossia; but it also vastly extends the domain of intersubjective relations.[12] The cyborg, which incorporates the mechanical and the animal, comes to exist in another relation to the animal and the mechanical. That animal and mechanical are no longer simply the "other" of the I/subject, but correspond to the "otherness" within the bastardized I/subject.

Xenotransplantation—the result of technological strategies of control—could have enormous consequences for the ways in which human beings involve themselves with animals. Haraway proposes that cyborgs, thanks to their contaminated character, possess the possibility (no guarantees, though) of such a denatured engagement—a kind of temporary coalition on the basis of affinity, not identity, with relations by choice, not blood (Haraway 1991, 155). Here again appears the responsibility of choice. At the same time this coalition can constitute an effective strategy of opposition: "I like to imagine . . . a kind of cyborg society, dedicated to realistically converting the laboratories that most fiercely embody and spew out the tools of technical apocalypse, and committed to building a political form that actually manages to hold together witches, engineers, elders, perverts, Christians, mothers, and Leninists long enough to disarm the state" (154–55). For Haraway, political action principally means forging the opposition.[13]

Ultimately the most crucial part of her thesis is that, once the radical constructibility and transformability of things, animals, human beings, and narratives is realized, their radical contingency, facticity, and partiality is disclosed. We should take up the challenge of affirming this situation rather than playing at murdered innocence; "surviving in the diaspora" without fear or regret may lead to a sometimes pleasurable and mostly risky way of life. This implies at the same time the challenge of developing oppositional strategies to dislocate the apocalyptic and totalizing narratives. This is possible, because along with the loss of innocence and the myth of original identity so, too, the urgency to identify the Enemy perishes.

4. Body, Machine, Text

Much of what Haraway says is not really new; similar ideas are found in postmodernist thought following the so-called linguistic turn. Human beings are understood as interpreters of signs, their identities are understood as collections of stories they tell about themselves, and things are the subjects of multiple stories. In this sense one can say that "the electron," "the neurosis," "the mind," and "the body" do not exist, although there do exist scientific, philosophical, and popular genres of narratives that have such things as their subjects, each of which possesses its own rules for meaningful discourse. Reality, in this view, is structured like language, concretized in many narratives and fixed in texts; knowledge becomes a form of textual interpretation and behavior a form of narrative telling or writing. This postmodernist approach, too, undermines naturalism, for it sees narratives as historical constructions and finds that there is no original text or "Book of Nature." Further reflection on this linguistic approach leads to the idea that texts are polyinterpretable, contaminated, and fragmented; that mutually incompatible genres can be mixed; and even that this is inevitable. Clearly, the "human condition" inside this textual universe coincides with what Haraway attributes to her cyborgs.

There is indeed a compelling parallel here, but there is still more at work in Haraway's thought. Cyborgs not only blur the boundary between human beings, animals, and machines, but also that between body and language. Modern technology textualizes reality, Haraway would say. This is clearly evident in information technology, a technology that is based on the manipulation of signs. But it is also evident in biological discourse about the body, which more and more tends to conceive of the body primarily as expressing or bearing the genetically coded information of DNA molecules. While for the eighteenth-century physician Luigi Galvani—who in 1786 brought a dead frog to "life," causing it to move by giving it a small electrical "kick"—electricity was the secret of life, today the secret of life is thought of as inscribed in a kind of text, which it is the aim of the Human Genome Project to decipher and store in a database. Life is conceived as a vast process of reading and writing, with ribosomes as the transcribers in their scriptorium-cell. Meanwhile, the life sciences have become a form of textual reading practice and cryptology; the technology of genetic recombination is a question of quoting one passage of a DNA text in another; genetic mutations are a form of textual corruption; and so on.

In the interesting final chapter of *Simians, Cyborgs, and Women,* entitled

"The Biopolitics of Postmodern Bodies: Constitutions of Self in Immune System Discourse," Haraway illustrates this textualization of the body using the example of immunology (1991, 203–30). The entire discourse of this branch of medical science is permeated by notions of coding and decoding, of identifying and misrecognizing information and "passwords." A virus, in this discourse, is a clever invader that sets out to produce an entirely new text that eventually is recognized as "foreign." An autoimmune disease is a structural misrecognition of a password—a disruption in communication. Immunology is about the immunological system, defending the body against invaders from the outside, which must therefore be able to establish the difference between friend and foe, health and sickness. Immunologists understand this as a process involving original texts, textual corruptions, and the ability to recognize alien texts as such.

But Haraway makes a still more important point. From her social-constructivist perspective she sees the immune system above all as an object in twentieth-century scientific narratives, and then asks herself what this narrative is saying. She claims, "[T]he immune system is a plan for meaningful action to construct and maintain the boundaries of what may count as self and other in the crucial realms of the normal and the pathological" (1991, 204). The ways in which the immune system figures in scientific narratives is as a defense mechanism that protects the "self" of the organism—the urtext—by closing it off against intrusion and corruption by an "other." This type of scientific narrative leaves little to the imagination; it is a war story, complete with all the intrigues, strategies, and betrayals that belong to that genre. In fact, Haraway claims, the entire discourse of original integrity, the necessity of exclusion, and the defense against the corrupting boundary breakdowns is repeated in the medical-biological narrative of sickness—the original narrative of the "self" legitimates the war against the "other."

But precisely this original narrative is problematic for cyborgs, for they, like OncoMouse™ are already corrupted mutants. And just as cyborgs have no need to defend their "original" identities against an "other," so their sickness stories have no need to be war narratives. Haraway provides an example drawn from Winograd and Flores's study "Computers and Rationality: The Myths and Realities" (1986) about the operation of computer systems in artificial intelligence. A central concept is communication *breakdown*, which Haraway then adopts, in her essay on immunology, as a metaphor for illness. Winograd and Flores conclude on the basis of their study: "Breakdowns play a central role in human understanding. A breakdown is not a negative situation to be avoided, but a situation of non-obviousness, in which some aspect of the network of tools that we are engaged in using is brought forth to visibility. . . . A breakdown reveals the nexus of relations necessary for us to accomplish our task. . . . This creates a clear objective for design—to anticipate the form of breakdowns and provide a space of possibilities for action when they occur" (in Haraway 1991, 214).

Haraway draws here a connection between the political (war) narrative and the prevailing immunological narrative. She claims that Winograd and Flores's interpretation points to a radically different kind of narrative than the prevailing one. "This is not a Star Wars or Strategic Computing Initiative relation to vulnerability, but neither does it deny therapeutic action. It insists on locating therapeutic, reconstructive action (and also theoretic understanding) in terms of situated purposes, not fantasies of the utterly defended self in a body as automated militarized factory, a kind of ultimate self as Robotic Battle Manager meeting the enemy (not-self) as it invades in the form of bits of foreign information threatening to take over the master control codes" (1991, 214–15). She adds that in these completely denaturalized systems another possibility for political action comes to light, quite different from that which was suggested by the military motives underlying their construction. This is precisely the irony of history to which cyborgs bear witness. And it is to be entrusted to responsible and tactful cyborgs to develop truthful narratives, such as that of Winograd and Flores. The construction or "writing" of such denaturalized narratives is one form of political action needed to disarm the state.

HOW RADICAL ARE CYBORGS?

In conclusion, a few passing remarks about Haraway's thinking. One could take her thought as one long sustained critique of naturalist foundational thinking. But in the end, how radical is her critique? Thanks to the shocking and subversive way she talks about cyborgs, her position seems to be much more radical than it actually is. This can be made clear via two questions.

First, one may ask how much genuine philosophical credibility there is in the "naturalism" that Haraway attacks. Naturalism, as she depicts it, is based on empirical criteria that can identify a boundary between the self or proper "nature" of something—of human being, man, woman, organism, machine, text—and the "other." This naturalism is thus a version of the "positivism of essences" or naive essentialism. Haraway is right to say that this figure of thought plays a large role in the mechanisms of exclusion and in political legitimations determining the Enemy—the one who exemplifies the alienation from that which counts as the proper nature. But one cannot say that cyborgs are required to undermine this figure of thought. Since the transcendental turn in philosophy at the end of the eighteenth century (Kant), and the hermeneutical-linguistic turn in this century, naive essentialism has been under sharp attack, and in fact it was philosophically stone-dead long before the first cyborg was constructed. But such philosophical outcomes do not have to affect political scenarios: naive essentialism may yet linger on in political strategies. And in the form it can take involving national and ethnic anxieties over loss of identity when one mixes with the "other," the obsession to determine enemies becomes alarmingly current. Philosophy is no match for such inclinations—but are cyborgs? There is reason to be skeptical. In my opinion, Haraway's cyborg thesis should not be conceived as an original philosophical reflection but rather as a politically emancipatory strategy. The originality of her thesis consists in the fact that she reproduces the critique of naive essentialism in the arena of technological development.

Second, one could ask whether cyborgs—assuming they blur the boundaries the way that Haraway says they do—have the political potency that she ascribes to them. What are we finally left with once we think through their boundary-blurring character radically and consistently? At first sight Haraway seems to undermine all fundamental distinctions between types of beings: distinctions between animal, woman, man, machine, kinds of texts, and so forth. However, in her own narrative one particular opposition is still left standing: living versus dead. Here is one distinction not yet overthrown. Haraway always dramatizes cyborgs as endowed with an (original) *conatus*, life force; the ambivalence of the cyborg consists in the different kinds of behavior that stem from this force. First, there is the cyborg as a project of militarist thinking, which serves "life" by being an actor in a defensive strategy in the battle of self (or one's own) versus life-threatening other. Second, there is the cyborg as contemplated by Haraway, which deconstructs precisely that image, which is in the service of surviving in the diaspora, and which militates against the obsessive aim for an "apocalyptic unity" and against equally life-threatening enemy images. Haraway has made naive essentialism her sworn Enemy; it stands for death and she combats it with a call to life.

Haraway claims that cyborgs stand in a different relation to death. Their surviving in the diaspora is characterized by situatedness, contingency, finitude, and vulnerability. They no longer know the dream of the total overcoming of all restrictions. Inside finitude they play the deadly serious game of surviving in the diaspora, the risky and sometimes carnivalesque construction of identities—all in the service of life. It may be that human beings can understand this lifestyle; it is questionable whether cyborgs can. If cyborgs are truly the boundary-blurring beings that Haraway represents them to be, then the breakdown of boundaries must be radically thought through and the possibility not be excluded that they also embody the blurring of the boundary between "living" and "dead." This would be the most monstrous irony: that ultimately in the battle for life versus death what becomes indistinct is the boundary between life and death. Cyborgs would then be not post-human but post-humous beings. The figure of the bionic man of science fiction literature would then flow readily into that of the undead Nosferatu in horror literature; a being that is partly alive, partly dead, or both together.[14]

Is it just chance that, today, the best representatives of Haraway's "self-regulating man-machine systems" are the half-alive, half-dead occupants of certain restricted areas in any medium-sized hospital? Thus far Haraway does not seem to have written about these. Possibly they fail to interest her because her project of obliterating "nature" is ultimately in the service of the effort to create a livable world for outsiders, and in that perspective she naturally thinks in terms of concrete human beings with a life to lose and therefore in need of a political strategy. But it would not be surprising if it turned out that cyborgs make very poor coalition partners, because beings that in the struggle against death ultimately blur the boundary between life and death are politically worthless. What could they reasonably strive for? There is no more apolitical genre than this kind of horror—and aren't cyborgs more at home in horror than in science fiction?

NOTES

1. Most of the material in the biographical section that follows is taken from this essay.

2. Jane Goodall and Dian Fossey were ethological field researchers. Fossey is above all known for her research into the behavior of mountain gorillas. She was murdered in 1985, presumably by a native poacher because of her efforts to protect "her" gorillas against the native population.

3. This George Orwell edition constituted part of the collected supplements to the journal *Das Argument: Blätter der Westberliner Studentengruppen gegen Atomrüsting*, which appeared in 1959.

4. Its title there is "A Cyborg Manifesto: Science, Technology, and Socialist Feminism in the Late Twentieth Century" (Haraway 1991, 149–81).

5. Cybernetics, from the Greek *kubernèsis*, means steering or the art of navigation. Today it refers to the technology of self-regulating devices and in a more general sense to information technology.

6. Haraway makes the comparison with chimeras explicit. One should also compare them with grotesques or pictures of hybrids of humans, animals, and plants. These bizarre figures are as much connected with the *unheimliche* or uncanny (a manifestation of the Freudian unconscious) as with carnivalistic vitality (e.g., Spek 1996, 84–85). Both characterizations also apply to cyborgs, which are at the same time monstrous and overflowing with life.

7. An integrated circuit is a technical term for a computer chip. Haraway uses it as

a metaphor for the proper connection of things in high-technological culture—just as the global village is a metaphor used to emphasize the falling away of literal distances.

8. This thought is explained in the next paragraph: according to Haraway the unity is always the result of a process of exclusion, so that the obtaining of the ultimate goal has the character of the ultimate conquest of an "own" over an "other" in a sort of Star Wars apocalypse: a deadly unity.

9. The concept of language game was coined by Ludwig Wittgenstein, who saw different kinds of language games occurring in different areas, including ethics, science, and personal expression. The criteria for the validity of an ethical expression ("thou shalt not kill") are of a completely different character from the criteria for the validity of a scientific expression ("iron expands when warmed"), and the criteria for each of these are in turn completely different from the criteria for the validity of a personal expression ("I love you"). Different language games therefore also involve different rules of argument, forms of rationality, and perspectives on reality. Ultimately, there are innumerable kinds of language games—praying, thanking, greeting, commanding, ordering, stating, and so forth. The universal language game, whose existence Wittgenstein denied, would offer an all-encompassing and universally valid perspective on all reality.

10. In the realm of literary representations, we are reminded here of the remarkable number of feminist authors of science fiction, who have transformed the "toys of the boys" genre into a literary laboratory in which to experiment with possible ways of existing in high technological culture (see Spek 1996). Haraway, whose writings can often be counted in that genre, sometimes refers to it explicitly (e.g., Haraway 1991, 173–81).

11. This situation is similar to that of a judge who must sometimes "construct" a sentence in the absence of a clear and unambiguous set of legal rules for dealing with the offense, while only partly being able to rely on historical precedent. Such a judge cannot be challenged for transgressing the rules, for these do not exist. The judge can, however, be challenged for his or her judgment. In this situation one can appeal to the classical *subtilitas applicandi*, to *Fingerspitzengefühl*, and to Aristotelian *sophrosunè;* the sort of practical wisdom or consideration that one gains only by living and learning; thus, only *after* innocence.

12. In her later work she frequently uses concepts such as "sibling" and "kinship" in order to affirm these intersubjective relations. She uses these concepts, of course, not to point to any "natural relationship" but rather to an awareness of conditions of connectedness that call for engagement.

13. Or perhaps better: practical oppositional activity, especially through the creation of new truths and narratives ("writing"). A model is given in the next paragraph on the basis of the immunological example.

14. Haraway's later work contains discussions of the undead, in the form of figures like Nosferatu, Count Dracula, and vampires (Haraway 1996, 74, 133, 152, 216, and 309 n. 3). But without exception these are trotted out as representations of the denatured "other" that are abhorred or pursued as a result of the obsession for the preservation of oneself or one's own. Haraway sympathizes with them, and would readily grant them a place among the living. She thereby overlooks the possibility that a being that obliterates the boundary between life and death could just as little appreciate a "place among the living" as a cyborg could the "Garden of Eden" (Haraway 1991, 151). But this last possibility disrupts Haraway's own political agenda.

BIBLIOGRAPHY

Bhavnani, Kum-Kum, and Donna Haraway. 1994. "Shifting the Subject: A Conversation between Kum-Kum Bhavnani and Donna Haraway." *Feminism and Psychology* 4, no. 4: 19–39.

Clynes, Manfred, and Nathan Kline. 1960. "Cyborgs and Space." *Astronautics* 13: 26–27, 74–75.

Haraway, Donna. 1976. *Crystals, Fabrics, and Fields: Metaphors of Organicism in Twentieth-Century Developmental Biology.* New Haven, Conn.: Yale University Press.

———. 1978a. "Animal Sociology and a Natural Economy of the Body Politic, part I: A Political Physiology of Dominance." *Signs* 4: 21–36.

———. 1978b. "Animal Sociology and a Natural Economy of the Body Politic, part II: The Past Is the Contested Zone: Human Nature and Theories of Production and Reproduction in Primate Behaviour Studies." *Signs* 4: 37–60.

———. 1979. "The Biological Enterprise: Sex, Mind, and Profit from Human Engineering to Sociobiology." *Radical History Review* 20: 37–70.

———. 1984. "Lieber Kyborg als Göttin! Für eine sozialistisch-feministische Unterwanderung der Gentechnologie." In *1984* (Argument-Sonderband 105), edited by Bernd-Peter Lange and Anna Maria Stuby, pp. 66–84. Berlin: Argument-Verlag.

———. 1985. "Manifesto for Cyborgs: Science, Technology, and Socialist Feminism in the 1980s." *Socialist Review,* no. 80: 65–108.

———. 1989. *Primate Visions: Gender, Race, and Nature in the World of Modern Science.* New York: Routledge.

———. 1991. *Simians, Cyborgs, and Women: The Reinvention of Nature.* New York: Routledge/Free Association Books.

———. 1994. "A Game of Cat's Cradle: Science Studies, Feminist Theory, Cultural Studies. *Configurations* 2, no. 1 (winter 1994): 59–71.

———. 1996. *Modest_Witness@Second_Millennium.FemaleMan©_Meets_OncoMouse™.* New York: Routledge.

Kuhn, Thomas. 1970. *The Structure of Scientific Revolutions.* 2d ed. Chicago: University of Chicago Press.

Minsky, Marvin. 1994. "Will Robots Inherit the Earth?" *Scientific American,* October, 86–91.

Spek, Inez van der. 1996. "A Momentary Taste of Being: Female Subjectivity, the Divine, and the Science Fiction of James Tiptree, Jr." Ph.D. diss., Rijksuniversiteit Utrecht, Enschede. Published as *Alien Plots: The Fiction of James Tiptree.* Liverpool: Liverpool University Press, 1999.

Winograd, Terry, and Fernando Flores. 1986. *Understanding Computers and Cognition: A New Foundation for Design.* Reading, Mass.: Addison-Wesley.

5. Don Ihde: The Technological Lifeworld

Peter-Paul Verbeek

What role does technology play in everyday human experience? How do technological artifacts affect people's existence and their relations with the world? And how do instruments produce and transform human knowledge? These are the central questions in Don Ihde's philosophy of technology.

Ihde, who was born in 1934, is a pioneer in two respects. First, he was one of the earliest philosophers in the United States to make technology the subject of philosophical reflection; and second, he was one of the first to apply to the study of technology the tools of the phenomenological tradition at a time when it was far out of the philosophical mainstream. Ihde studied theology with Paul Tillich, in the course of which he became interested in the philosophy of existence of Heidegger and Jaspers. He graduated from Boston University in 1964 with a thesis on the philosophy of Paul Ricoeur, which was the first systematic study in English of Ricoeur's work. Ihde continued to publish in the area of phenomenology, and in the early 1970s discovered and began writing essays on technology as an area of phenomenological exploration. He published his first book on the philosophy of technology, *Technics and Praxis,* in 1979, the first of over half a dozen books he has written in that field. He also began, and continues to edit himself, a book series devoted to the philosophy of technology, The Indiana Series in the Philosophy of Technology. Ihde's most important book, *Technology and the Lifeworld,* appeared in that series in 1990. This book draws together systematically the most important elements of his philosophy of technology.

Like the other philosophers discussed in this volume, Ihde develops a new perspective on technology, a perspective that seeks closer contact with concrete technologies. Classical philosophy of technology tended to reify technology, treating it as a monolithic force, "Technology." Ihde, by contrast, shuns general pronouncements about Technology, fearing to lose contact with the role that concrete technologies play in our culture and in people's everyday lives. He sets himself the task of exploring this very role of technologies. Ihde does so from within the phenomenological tradition, which he has helped to connect to the philosophy of technology. His work departs from Heideggerian-style tradi-tional phenomenological analyses of technology, and redirects that tradition in new ways with respect to the philosophy of technology.

I shall begin this chapter with a short description of phenomenology as a philosophical method, followed by a review of some key elements of Heideg-ger's analysis of technology, in order to create a background against which Ihde's philosophy of technology can be introduced. My account of Ihde's work will consist of three parts. The first will concern Ihde's analysis of the role of technology in human experience, the second his vision of the relation be-tween technology and culture, and the third the implications of his philosophy of technology for the philosophy of science.

I. PHENOMENOLOGY AND THE PHILOSOPHY OF TECHNOLOGY

The phenomenological approach has always occupied an important place in the philosophy of technology thanks in large measure to the great influence of Heidegger, who devoted considerable attention to technology. Phenom-enology is a philosophical approach that seeks to overcome the classical, Cartesian dichotomy between subject and object. Against this dualistic no-tion, phenomenology holds that subject and object—in later phenomenology this becomes "human beings" and "world"—cannot be thought independently of each other, but only as always already related. Humans cannot be conceived apart from their relations to the world, and the world cannot be conceived apart from people's relations to it.

In phenomenological terminology, this connectedness of humans and world is called *intentionality*. This intentionality must not be seen as a goal-conscious activity of human beings. This would imply that the relation of hu-mans and world comes about through a voluntary act that they could just as well refrain from doing. Such a vision again would amount to a sharp division between humans and world, whereas humans are instead related to the world whether they want it or not—they cannot be otherwise. Human consciousness can never be adequately understood or described in isolation as consciousness-in-itself. It never exists by itself, but only as consciousness-of-something. The same is true of perception: there is no perception-in-itself, since perception is always perception-of-something. Conversely, in the phenomenological perspective, there is no such thing as a thing-in-itself, either. If we experience things, we experience things-for-us, that is, things as they are revealed to us. Humans have no direct access to reality. Their reality is always what it is because it is revealed to them within a specific context of interpretation or praxis. Once again, the same goes for perception; just as perception is always a perception-of, it is also always a perception-as. To relate to the world is to interpret it. We never find ourselves in "the" world but always in "our" world. Things are what they are by virtue of our relations to them, just as we are what we are in terms of our relations to things. Humans and world are inseparably bound to each other and constitute each other in this bondage.

What, then, does all of this have to do with technology? German philosopher Martin Heidegger was one of the first to think about technology from this phenomenological perspective, and his views are an important part of the background against which Ihde's philosophy of technology is to be understood. Central to Heidegger's thought is the question of "being" as the ground of the continuing underlying relations of humans and their world. Humans can have a relation to beings—that is, to everything that "is"—only by virtue of the fact that those beings "are." This "being" of beings always involves a moment of transcendence, of not being reducible to us. That beings "are" and can be disclosed in our relations with them, is not our doing, but a situation in which we always already find ourselves. The "being" of beings transcends humans: it is not their product, but the omnipresent horizon of their relation to beings.

According to Heidegger, "being" must be conceived historically. Being, in Heidegger's terms, "sends itself"—*schickt sich*—in different ways in different historical epochs. In each epoch, "to be" means something different, which leads to a different disclosing of the world. This does not imply, however, that "being" must be conceived as determined by humans. Humans have only limited ways to determine how their world is disclosed, for that disclosure always happens against the background of a particular and historical meaning of "being." This background is not produced by humans, but is something in which they always already find themselves. Being has its own history, which cannot be completely understood through human interventions.

According to Heidegger, modern technology must be understood not as the sum total of technological devices, but as the ways in which our present-day epoch discloses the world. The most distinctive and essential feature of the technological way of world-disclosure is that it involves a forgetting of the moment of transcendence, which is indissolubly related to the being of beings. Technology is domination and control; it "sets upon" or "challenges" nature; it is a process involving the appearance of nothing transcendent, nothing beyond beings. In the technological form of the disclosure of reality, the world, according to Heidegger, is a *Bestand*—a stock or warehouse—a collection of goods that have been stored up and made ready for manipulation and use. The technological disclosure of the world is not a "letting be" of the world but rather a "summoning" or "enframing" (*bestellen*) of it. What counts as reality is what can be made and manipulated. In modern technology, we encounter only beings, and no longer the "being" that is their ground. We have put ourselves in the place of that ground. In Heidegger's words, we have "forgotten" being.

Heidegger, therefore, thinks of technology as a way—but only *one* way—in which reality can be disclosed. Insofar as he thinks of concrete technological artifacts, it is as the products of this technological interpretation of reality. Heidegger does this deliberately, for his project is to understand "the essence" of modern technology. According to him, "the essence of technology is by no means anything technological," any more than the essence of a tree can be found among trees (Heidegger 1977, 4). Heidegger wants to understand technology as more than "a means to an end" or "a human activity." Technology, according to him, *is* indeed a means for ends and a human activity—these are what he calls the instrumental and anthropological definitions of technology—but its "essence" lies at a deeper level. Heidegger wants to understand "within what do such things as means and end belong?" (6). He thus questions "backwards," from concrete equipment and procedures to the underlying way in which reality is disclosed, which makes these equipment and procedures possible. In short, his questioning is not about "technologies" but about "Technology."

This approach was firmly rooted in the tradition of the philosophical thinking about technology in its early days. The pioneers of the philosophy of technology generally occupied themselves with trying to identify the essential characteristics of technology, and with trying to clarify its role in our culture. But the new generation of thinkers about technology discussed in this book no longer thinks in terms of Technology per se, and finds it problematic to try to understand phenomena in terms of essences. The philosophers of this new generation adopt various approaches in concerning themselves instead with concrete technologies and the roles they play in their specific contexts.

Don Ihde's approach, as mentioned, accomplishes this turn from Technology to technologies within phenomenology. In his thinking he breaks with the phenomenological tradition's conception of technology as stemming from a specific and limited way of disclosing reality. Ihde seeks to reflect about technology as it is concretely present in our daily existence: in the form of technological artifacts. Instead of questioning "backwards" he questions "forwards";

that is, instead of reducing technological artifacts to the technological form of world-disclosure that makes them possible, he asks what form of world-disclosure is *made possible by* technological artifacts.

The scope of Ihde's attention therefore is the relation of human beings to technological artifacts. From the perspective of human-technology relations he tries to understand the role of technology in the human lifeworld. Ihde inquires into the relation between human beings and technological artifacts on two levels: experience and culture. On the level of experience he inquires into the role that technological artifacts can play in the relation of human beings to reality; on the cultural level he inquires into the relation between technological artifacts and culture. These two sides of Ihde's research in the philosophy of technology will be discussed separately, followed by an overview of the way Ihde connects his philosophy of technology to the philosophy of science.

2. HUMAN-TECHNOLOGY RELATIONS

Ihde's conception of phenomenology is a manner of thinking that occupies itself principally with human experience, and specifically with the *structure* of experience (Ihde 1990, 21, 23; 1986, 21). Ihde calls his analysis of human experience *relativistic* (1990, 23–25; 1998, 46)—not in the sense of an epistemological relativism, but rather in the more literal sense of an *analysis of relations:* "A phenomenological account . . . always takes as its primitive the relationality of the human experiencer to the field of experience. In this sense, it is rigorously relativistic. The relationality of human-world relationships is claimed by phenomenologists to be an ontological feature of all knowledge, all experience" (1990, 25).

It is understandable that, according to Ihde, experience plays such a crucial role in phenomenology, since experience is the place where the mutual relation between human beings and their world can be localized. Ihde analyses human experience in terms of perception. He considers perception the key to understanding what was just called the "relation between human beings and their world." Perception is as it were the interweaving of both: in perception, human beings and world—or subject and object, for that matter—are not separated but always intertwined. Only afterward, when a perception is described and not enacted, does it make sense to separate out a perceiver and a perceived; or a subject and an object, as one says. In the perceiving itself that cannot be done, since to perceive is to perceive *the world.* In experiencing, people are as much "in" the world as the world is "in" them: they cannot be separated.

Ihde distinguishes two dimensions of perception. The first is sensory perception, a bodily dimension that Ihde calls *microperception*. The second, an interpretive dimension that discloses meaning and is cultural in nature, he calls *macroperception:*

> What is usually taken as sensory perception (what is immediate and focused bodily in actual seeing, hearing, etc.), I shall call microperception. But there is also what might be called a cultural, or hermeneutic, perception, which I shall call macroperception. Both belong equally to the lifeworld. And both dimensions of perception are closely linked and intertwined. There is no microperception (sensory-bodily) without its location within a field of macroperception and no macroperception without its microperceptual foci. (1990, 29)

While it is true that microperception and macroperception can be distinguished from each other, they cannot be separated. A bodily perception can no more exist without being interpreted than an interpretation can exist without something to be interpreted. The two-fold meaning of the word *perception* to which Ihde points is illustratively present in the verb "to see," which we can use to describe a bodily-sensory perception ("I see a tree") and to characterize an interpretation of the world ("Since that talk I see things completely differently").

Ihde pursues his analysis of the role of technology in the interrelation of human beings and world by inquiring into the forms of these interrelations when technological artifacts are involved. To that end he distinguishes three different ways in which human beings can relate to technological artifacts. The first of these human-technology relations is the relation in which our perception is mediated by a technological artifact. In such a "relation of mediation" we are not *directly* related to the world but *via an artifact*—as for instance whenever we wear glasses or watch television. A second kind of relation, which Ihde calls an *alterity relation,* is a relation not via an artifact to the world but to an artifact itself. The third kind of human-technology relation Ihde calls a *background relation,* in which technological artifacts shape our relation to reality but by remaining in the background, as do thermostats that automatically switch the heat on and off without our intervention or even awareness. These human-technology relations will be discussed separately below.

Relations of Mediation

Technologies play an important role in our daily lives by mediating our experience. We read off the temperature via thermometers, we remember events via photographs, we speak with each other via telephones, and we keep abreast of current events via television. In all of these cases we are not directly in bodily-sensory experience present to the world but via technological artifacts. What happens to our perception when it is enacted via technologies?

Ihde begins to work out his answer to this question with the aid of Heidegger's analysis of tools and Merleau-Ponty's analysis of the role of "embodiment" in perception. Heidegger, in his analysis of equipment, asks himself how best to characterize the way that a tool or useful thing is present to human beings. He concludes that a tool is "something in order to"; it is serviceable, helpful, usable (Heidegger 1996, 64). A tool does not exist by itself, but in a context to which it refers. A hammer is for hammering: it is most present to us as a hammer when we hammer with it rather than use it as a paperweight, for instance. More generally, the way to encounter a tool as a tool, and not just as an object lying around, is not to examine it theoretically, but to take it up and actively use it. When this happens, the tool has the kind of being that Heidegger calls "handiness" or "readiness-to-hand" (*zuhandenheit*). It is characteristic of something handy that it withdraws itself in order to be handy. Someone who is hammering is not concerned with the hammer but rather with what is being done, or made, with the hammer. Only when human beings are not concerned with the tool but rather with the work that they are doing with their tools do these tools become present to them *as* tools. Tools call attention to themselves only if and when it is impossible to do anything with them; we notice our handy tool-objects only when, for whatever reason, they suddenly become unusable. The reliable dealings we are accustomed to having with them is disrupted, and the object suddenly forces itself on us, as does a hammer when its head flies off the handle. The tool is then "objectively present" or "present-at-hand" rather than "handy."

Ihde finds in Heidegger's analysis of the ways in which tools are present for human beings three elements of special significance. First, Heidegger shows that each tool, each piece of equipment is related to a context. In itself it is nothing; as a piece of equipment it is a part of a meaningful whole. "This field within which a tool is what it can be is a complex one filled with 'involvements' or cross-relations" (Ihde 1990, 32). Second, it is clear from Heidegger's analysis that equipment has an "instrumental intentionality"; a tool is "something in order to," and in that "in order to" there is always a reference of that tool to a context, to whatever can be done with it. Later in this chapter we shall return to this concept of "technological intentionality." The third element Ihde considers important is that Heidegger shows that the tool, when used in practical activity (and not in descriptions of it), is a means of experiencing, rather than an object of experience.

This last element is central to Merleau-Ponty's analysis of the role that objects can play for human beings. Ihde finds in Merleau-Ponty a still deeper sensitivity to the relation between the human body and the world that lays the groundwork for a phenomenology of technology, and points to two examples that Merleau-Ponty gives that are especially illuminating concerning how human beings are related "through objects" to the world: "the woman with the feather in her hat" and "the blind man with the cane." Merleau-Ponty uses these examples to show that human beings can use artifacts to stretch the spatiality of their bodies. A woman with a feather in her hat can extend her area of sensitivity to the world to the point where she can keep a safe distance between the feather and objects that might damage the feather, stooping instinctively when necessary; "she feels where the feather is just as we feel where our hand is" (Merleau-Ponty 1962, 143). The image of the blind man's cane carries this a step further, making it clear that human beings can not only extend the spatiality of their lived bodies with the aid of artifacts but perceive with them as well. With his hand, a blind man feels not so much the stick as the street and the objects in the way *through* the stick. Just as Heidegger's carpenter is not involved so much with the hammer while at work as with the nail to be nailed in place, so a blind person is not truly involved with the cane as with the world through the cane. These images of Heidegger and Merleau-Ponty are in effect complementary. While Heidegger analyzes the ways in which artifacts are present to human beings, "withdrawing" from their experience, Merleau-Ponty analyzes the relations to the world that can arise on the basis of this presence.

These analyses, taken together, point to a structure of perception that can be described in terms of mediation. The intentional relation between human beings and world is as it were extended or stretched out through artifacts. Ihde schematizes the difference between unmediated and mediated perception as follows:

unmediated perception: I—world
mediated perception: I—technology—world

It must be noted that by "unmediated" Ihde means unmediated *by artifacts*. As will shortly be seen in the discussion of Ihde's conception of hermeneutics, all perceptions are in a certain sense mediated, because human beings never have direct access to the world but only via interpretation. Ihde is not concerned here with mediation of this type—through language, for instance. When he speaks of "naked perception," he means not some pre-interpretive access to reality but a perception that takes place without the intervention of an artifact on the microperceptual level.

In analyzing a number of examples of mediated perception, Ihde comes to the conclusion that there are two basic sets of relations with artifacts in which they mediate people's relations with their world. The first involves what he calls *embodiment relations*. In these relations, human beings take technological artifacts into their experiencing, and thereby broaden the area of sensitivity of their bodies to the world. An example of the embodiment relation involves the wearing of eyeglasses. When I wear eyeglasses, I do not look at them but through them at the world. I take as it were the pair of glasses into myself; it withdraws from my perceiving. But embodiment relations are not restricted to the visual. A dentist, for example, who uses a dental probe to feel out cavities in my teeth is using the probe to extend the sensitivity of touch, feeling cavities via the probe. Ihde schematizes embodied relations as follows:

embodiment relations: (I-technology) → world

The most important characteristic of embodied technologies is that they possess a certain transparency. They call attention not to themselves, but to (aspects of) the world given through them. In order for this transparency to occur, however, several conditions must be met: (1) The artifact must be technically serviceable; that is, its physical characteristics must allow it to be embodied. A pair of glasses made with opaque glass cannot serve embodied perception. (2) A certain skill or technique is required to perceive through the artifact; those not trained in dentistry cannot use dental probes to detect tooth decay. (3) The artifact should aim at making mediated perception take place in a way comparable to unmediated perception; a telescope ordinarily delivers a picture of a planet with roughly the same size as a microscope ordinarily delivers a picture of a red blood corpuscle—"the image size of galaxy or amoeba is the same" (Ihde 1990, 79).

The second set of mediated relations with artifacts consists of *hermeneutic relations*. In hermeneutic relations, too, we are involved with the world via an artifact, but the artifact is not transparent. The artifact does not withdraw from our relation to the world but provides a representation of the world, which requires interpretation in order to impart something to us about it. Because this relation involves interpretation (the artifact must be "read"), Ihde calls it hermeneutic, using the traditional term for the philosophical-theological discipline of reading signs—though his usage, as we shall see, is somewhat unconventional and provocative. In hermeneutic relations the world is not perceived through the artifact but by means of it. Ihde schematizes hermeneutic relations as follows:

hermeneutic relations: I → (technology-world)

An example of a hermeneutic relation with an artifact is the use of a thermometer. When we read a thermometer, we are not involved with the thermometer itself but with the world of which the thermometer reveals an aspect: its temperature. This revelation, however, does not have the character of a *sensing* of temperature but is rather a *representation* of it.

So what are the implications of these technological mediations for our experience? Mediation, for Ihde, is indissolubly linked with a transformation of perception. Naked perception and perception via artifacts are never completely identical. In this transformational character of technological mediation lies an important aspect of the non-neutrality of technology. This transformation of perception has, according to Ihde, an extensive structure involving amplification and reduction. Mediation always strengthens certain specific aspects of the reality perceived and weakens others. Whenever we look through a spyglass, for instance, we see objects that we wouldn't otherwise, and to this extent our visual access to reality is strengthened. But at the same time, we do not hear, smell, or feel what we see; our perception has been reduced to the visual. The dental probe is still a better example, since it shows the structure of amplification and reduction within the same sensorial field: it improves certain kinds of feeling while curtailing others.

The transformation of perception, with its structure of amplification and reduction, appears in different gradations. When we compare mediated perception with naked perception, we can distinguish between transformations of low contrast and transformations of high contrast with respect to perception "with the naked eye." The transformation that a pair of eyeglasses brings about, for instance, is a transformation of low contrast. The world that eyeglass wearers perceive strongly resembles the world that they saw before they needed glasses; the only difference is that the image is enframed. The transformation effected in a spectrogram, however, is of an entirely different order. A spectrogram is a visual deposition of the light given off by, say, a star, from which information can be derived about the star's chemical composition. The spectrogram's band of colored stripes is as removed as possible from the star that we see with the naked eye; nevertheless, it reveals important aspects of the star. This is a transformation of high contrast.

These mediations of microperception have consequences for macroperception, for the manner in which human beings interpret their world. Embodiment relations and hermeneutic relations can be viewed as the extremes of a continuum. As we move on this continuum of embodiment to hermeneutic relations more toward the hermeneutic pole, the transformation that reality undergoes in the mediation is one of progressively higher contrast: the perception effected by the mediation deviates ever more sharply from unmediated perception. The reason for this is that what mediated perception can make visible is determined with ever more specificity as the mediation grows more hermeneutic in nature. A hermeneutic technology, after all, provides a representation of reality, which implies that the design of such a technology predetermines which aspect of reality is to be made perceptible by it and in which ways. The "space" available for reality to express itself becomes more restricted as the mediation of our perception becomes more hermeneutic in nature.

A pair of eyeglasses, for instance, that effects a transformation of extremely low contrast, provides access to reality in practically the same domain and with the same possibilities of interpretation as perception that is not mediated by technology. A spyglass and a telescope do that to a lesser extent, in that some of "the whole" of experience available in unmediated perception must be given up: experience is reduced to vision here. At the same time these technologies open worlds that previously had been hidden, since they allow us to see things we could not see without them. The transformation effected by a microscope is still greater because it makes perceptible a reality that deviates more strongly from our daily reality. Although biologists and medical specialists are able to embody the microscope in such a way that they know their way about in the new reality it discloses, this reality differs strongly from the world in which we live: looking in a microscope puts you in another world, which is more difficult to describe in terms of our everyday lifeworld. A spectrogram, finally, reveals reality only in terms of scientific phenomena, which further restricts the number of possible interpretations. It reveals only one aspect of reality, and a scientific one at that: the chemical constituents of the phenomena perceived.

The insight that technologies can play a mediating role in our experience, in which certain aspects of the world are strengthened and others weakened, therefore points to the need to nuance the classical, Heideggerian thought that technology consists of a specific, and reduced, interpretation of the world. According to Heidegger, technology is characterized by an interpretation of the world as "standing reserve," or *Bestand;* as a storehouse of goods that lie ready for human manipulation. This interpretation allows the world to appear only in a very limited respect: under the guise of control and domination. But whenever we consider technology in terms of concrete, mediating artifacts, as does Ihde, it becomes clear that our dealings with these artifacts do not require us to have such a "controlling" interpretation of the world. A tree is not forced to show itself as firewood or as potential furniture material when viewed through a pair of eyeglasses; rather, the pair of eyeglasses opens up to its wearer the same domain of possibilities of interpretation as are available to the non-eyeglass wearer. But there is more: technology can even allow the world to manifest itself in new ways. Ihde points to infrared photography as an example. In this form of photography we lose, to be sure, the non-visual aspects of the photographic object and the depth of the unmediated image, but at the same time it makes perceivable things that remain invisible to the naked eye (Ihde 1991, 73–74). It is easier to tell whether the trees are diseased on the basis of infrared photographs of trees than through inspection with the naked eye.

Technology, conceived as concrete technological artifacts, therefore does not necessarily reduce our relation to the world to "setting upon" or "summoning." It can indeed constrict our access to the world, but at the same time it offers us different ways of access to the world, even ones that would be impossible without technology. The nature of these ways of access varies from technology to technology, because technologies transform perception differently. Technologies are more ambivalent than alienating, with respect to the interpretations of the world with which they are linked. When they mediate our experience, they have as much a reductive as a strengthening impact on our experience. The more it is possible to embody a technology, the less it predetermines in which ways the world can manifest itself through it, and the less it reduces our interpretation of the world. Moreover, both embodied and hermeneutic technologies can make possible new modes of access to the world, which would be impossible without mediation.

One question that might arise from this analysis of technological mediation is whether Ihde remains faithful here to the phenomenological ambition of overcoming the subject-object schema. For he appears to affirm that schema in his analyses of the roles of technologies in the relations between humans on the one side and the world on the other. By locating mediation "between" human beings and world ("I-technology-world"), Ihde seems to put subject and object over against one another, instead of starting from the idea that they mutually constitute each other. This problem deserves some attention, since it gnaws at the roots of this approach to the phenomenology of technology.

There is a solution to this problem. The central thought of phenomenology, that subject and object must be thought of as mutually interwoven, does not necessarily clash with Ihde's analysis of technological mediation. It might be tempting to conceive mediation as a process in which a transformation occurs of the manner in which a subject (human) experiences an object (world)—in other words, as a process between a fixed subject and a fixed object in which only the manner in which the object is experienced by the subject is affected. Yet from a phenomenological point of view this is not what is happening in technological mediation. For a phenomenologist, the interrelation between subject and object always precedes the subject and the object themselves; the subject and the object are constituted in their interrelation. This notion of mutual constitution must be borne in mind when considering Ihde's discussion of the various relations between humans and artifacts. The difficulty with this, however, is that the "interrelation" of subject and object—or human and world—concerns a level that in fact precedes subject and object (human and world), and that there is no way to speak about this interrelation without making use of the words "subject" and "object," or "humans" and "world."

Mediation by artifacts must, therefore, not be seen as a mediation "between" subject and object, but as a mediation of a sort in which both subject and object are constituted. Mediating artifacts shape not only the way a predefined subject relates to a predefined object or the way a predefined object can appear to a predefined subject. They shape the interrelation itself between subject and object, from which both are constituted. Mediation does not simply take place *between* a subject and an object, but rather *co-shapes* subjectivity and objectivity. Formulations in terms of the "access to reality" offered by an artifact should be read as relating to the way in which an artifact makes possible the constitution of a world and a human in the very process of perception. Humans and the world they experience are the *products* of technological mediation, and not just the poles between which the mediation plays itself out.

Alterity Relations

The second human-technology relation described by Ihde is the alterity relation. In alterity relations humans are not related, as in mediating relations, via a technology to the world; rather, they are related to or with a technology. The role played by technologies in this set of relations can be characterized as that of a "quasi-other." This set of relations can be formalized as follows:

alterity relations: I → technology (-world)

Technology appears in alterity relations as quasi-other, because while we may encounter technologies in ways in which they seem to behave as an "other," they can, of course, never be present as a true person. Humans often approach the technologies that they encounter in anthropomorphic ways: they project human properties on artifacts ("intelligent computers"), or entertain certain feelings for them ("caring for" a piano). As Ihde points out, an automobile can indeed be seen as an other that I can care for, but it is far less of an other even than a horse, in whose place I can also put myself, but which does not always obey—and can even start or rear if a rabbit happens to cross its path. The otherness of technologies is of a completely different sort. The reason that technologies in alterity relations are experienced as quasi-other is that technologies on the one hand possess a kind of independence and on the other hand can give rise to an "interaction" between humans and technologies. Many toys, such as tops and music boxes, are fascinating precisely because of the apparent autonomy that they possess. Robots possess such an autonomy, to the extent that one could truly speak of "interacting" with these technological "beings." And today, automatic train ticket machines can not only take money and dispense tickets, but also give advice, provide route information, answer questions, and protest when something is done incorrectly.

As mentioned above, hermeneutic relations and embodiment relations form the extremes of a continuum. That continuum, however, is but a part of a still greater continuum, in which alterity relations also play a role. On the one end of this continuum are embodiment relations, in which technologies have the role of a quasi-I. In embodied relations technology always coincides as it were with myself. At the other extreme are alterity relations, in which technologies are present as a quasi-other, as indicated above. Between these two are hermeneutic relations, in which the technology on the one hand mediates and is therefore not present "as itself," but at the same time draws attention to itself because it is not embodied but "read."

Background Relations

The final set of human-technology relations that Ihde identifies is background relations. In contrast with the two kinds of relations already discussed, technological artifacts in background relations do not play a central role in our experience. In background relations, we are related not explicitly to a technology or via a technology to the world; instead, technologies shape the context of our experience in a way that is not consciously experienced. Schematically:

Background relations: I (-technology/world)

Refrigerators and central heating systems are examples of technologies with which we can have a background relation. These technologies switch themselves on and off in the background of our experience; we notice scarcely if at all that the room temperature is almost always the same and that the refrigerator is on. Technologies give rise to a background "field" in which we can have experiences without explicitly experiencing these technologies. They are present and absent at the same time: without us noticing them, they give form to our experience by shaping a context for it. And they can have many of the same transformational characteristics as the other technologies mentioned, with which our involvements are much more explicit. Their (absent) presence is usually experienced only when they stop functioning—when a storm knocks out the electricity, for instance. In such cases the context shaped by the background technologies, which we otherwise take for granted, is suddenly not self-evident any more.

3. TECHNOLOGY AND CULTURE

Ihde seeks to understand human-technology relations not only on the level of human experience, but also on the level of culture. Clarifying the relation between technology and culture has been the ambition of many philosophers of technology. In all the various approaches to this relation, two extreme positions can be characterized, which can, after Borgmann, be called *instrumentalism* and *substantivism* (cf. Borgmann 1984, 7–12). Instrumentalists see technology as a mere tool, as the means to accomplish certain tasks. Within this approach technology appears as something neutral; if technology is viewed as a means to an end, this implies that it is not to be judged in itself but only in its use. Substantivists, on the other hand, think that instrumentalists overlook the fact that technology is not at all neutral. According to the substantivists, technology must be understood as an independent power that can alter culture drastically. The substantivists attribute two properties to technology. First, they conceive of technological development as something autonomous. Technology always gives rise to new possibilities, and these new possibilities are always realized eventually. Nothing can stop technological development; technology follows its own dynamic. Second, the substantivists ascribe to technology the ability to change culture. Technological development, so to speak, takes culture along with it.

Ihde considers both positions to be unsatisfactory. Underlying both instrumentalism and substantivism is an assumption that he finds to be untenable: that one can speak about technology independently of the humans who are involved with it and the culture in which it functions. Technologies, according to Ihde, do not exist "in themselves," but only as related to humans and to culture, for humans are always and only involved with technology in a cultural context. One could say that Ihde is making here the same phenomenological move that others in that tradition made with respect to "consciousness" and "perception." Just as perception-in-itself and consciousness-in-itself do not exist, neither does technology-in-itself. Just as perception can be understood intentionally only as perception-of, and consciousness only as consciousness-of, so technology can only be understood as technology-in-order-to. The "in order to" indicates that technologies always and only function in concrete, praxical contexts and cannot be technologies apart from such contexts. In Ihde's words: "Were technologies merely objects totally divorced from human praxis, they would be so much 'junk' lying about. Once taken into praxis one can speak not of technologies 'in themselves,' but as the active relational pair, human-technology" (Ihde 1993, 34).

But just as technology cannot be grasped in isolation, neither can culture. A culture is only what it is in the praxes in which it manifests itself—praxes that are mediated by technologies. Not only does technology become what it is in and through the interweaving of technology and culture, so does culture and the human beings using the technology: "The human with a steel axe is different than the human without one—the transformational effect becomes

Figure I The Necker cube.

clear when we regard as the *primitive* of our analyses, this *human–technology pairing*" (Ihde 1993b, 34).

Multistability

The insight that technologies are indissolubly linked with humans-in-culture implies that technologies have no "essence"; they are only what they are in their use. Ihde names this ambiguity of technology *multistability,* and to clarify what he means he makes use of a perceptual example, the so-called Necker cube (figure 1).

When we look at this figure, we can see more than one thing. Sometimes we see a three-dimensional cube with the top surface and two side surfaces facing us, while at other times we see a cube with the bottom surface and two side surfaces turned toward us. If we try, we can make what we see switch between the two cubes. We can also interpret the figure two-dimensionally and see it as an insect with six legs sitting in a six-sided cell of its web. Ihde uses this example to illustrate that different ways of seeing produce different figures. The figure allows multiple interpretations. What it "really" is remains undetermined. It is many things at once; it is "stable" in multiple ways.

Something similar, according to Ihde, is at work in the relation between culture and technology. As with the Necker cube, so with technologies: one cannot say in isolation what they "really" are. Technologies are always technologies-in-use, and this use context is part of a larger cultural context. This contextuality makes technologies multistable, in a way that is analogous to the different possible ways of "seeing" the Necker cube: the same artifact can have different meanings or identities in different cultural contexts. Ihde illustrates this multistability with the aid of examples involving technology transfer between cultures. One of these examples concerns the "cultural embeddedness" of sardine cans in New Guinea, which in the 1930s were left behind by Australian gold prospectors and snatched up by the inhabitants for use as the centerpieces of their headgear (Ihde 1990, 125). But Ihde also uses examples closer to home; the early development of the typewriter and the telephone, he says, were driven by the desire to design equipment for the blind and hard of hearing to help them hear and write. The context in which they actually functioned, however, quickly defined these devices in a way that was much more than what they were designed for. They were seen not only as aids for the blind and hard of hearing, but as devices that were meaningful and useful for everyone (Ihde 1993a, 116). Feenberg's analysis of the introduction of Minitel in France, discussed in chapter 3, is another good illustration of the multistability of technologies. What these technologies "really" are cannot be determined with finality.

The multistability of cultural relations to the world implies not only that artifacts can have different meanings in different cultural contexts, but also that the same goals can be technologically realized in different ways. Ihde's favorite example of this is the difference between Western navigational techniques and the traditional navigational techniques of the South Sea islanders. While Western navigation is strongly instrumentally mediated and mathematical in nature—one navigates with charts and compasses—the South Sea islanders navigate by carefully observing stationary clouds (which hang over islands), birds, and wave patterns (Ihde 1990, 146–49). The South Sea islanders had an extremely complex navigational system and could navigate at least as well as the first Westerners who encountered them at the time of the first voyages of discovery. Though their navigational system was not technological in nature, instruments could in principle play a role in them, in the form of instruments that could peer through mist, make wave patterns more perceptible, and so forth. The cultural "way of seeing" of the South Sea islanders could then give rise to an entirely different technology than the Western one involving charts and compasses. The South Sea system always "looks" laterally from the position of the navigator, which calls for a completely different type of navigation than the Western system, which looks from overhead, down on the water and land. Human ends, therefore, can be realized in many different ways, depending on the cultural context in which they play a role. Different cultural contexts, different "ways of seeing," thus can lead to the development of different technologies.

The insight that technology can be described only as interwoven with culture, and that this interwovenness makes technologies multistable, makes the substantivist position untenable. Technology cannot be understood as an independent power that holds culture in its grip, for its form is ambiguous; it becomes what it is only in the context of culture.

Technological Intentionality

Following this discussion revealing the cultural index of technology, we seem to be back at the instrumentalist position where we began. If the cultural context determines what a technology is, it seems to follow that the technology itself has a certain indifference with respect to what it is, and therefore cannot play any significant role in culture and everyday life. But this is not the case. It is true that Ihde's remarks about multistability are intended to undermine substantivist conceptions that reify technology into a unified and stubborn force. Nevertheless, his deconstruction of this determinism does not at all turn technology into something soft and pliable that assumes whatever meanings culture would give it. Technology, according to Ihde, possesses a certain "robustness" and therefore is as little neutral as it is determining. Ihde calls this robustness of technology *technological intentionality* (Ihde 1990, 141). By this he means that because technologies provide a framework for human actions, they have a certain influence on those actions.

This influence does not have the character of a determinism but rather that of an inclination or "trajectory." Technologies "want" people to do things in particular ways, as it were: they have a certain "intention" and promote this intention among their users.[1] As an example, Ihde mentions the difference in writing style that arises when one writes with a fountain pen, typewriter, or word processor. One writes slowly with a fountain pen, with the result that it allows one to think over the sentence several times while composing it. The compositional speed is much faster with a typewriter, which tends to promote a style much closer to that of spoken language. And a word processor, in contrast to pen and typewriter, vastly expands the ability to compose a text; for instance, sentences can be moved around and footnotes inserted at will. These writing technologies do not have a determining influence, for one can indeed write a slowly composed and carefully thought out text on a word processor and can capture the cadence of spoken language using a pen. But the technologies in question incline toward a distinct use. The thoughts that Ihde develops here concerning "technological intentionality" recall Winner's discussion of the "politics of artifacts" and what constructivists call the "script" of technological artifacts. Technologies, so to speak, can play a role of their own when people use them. Far from being neutral, they can "ask" in compelling ways for specific ways of being used; they can contain their own implicit "application manual."[2]

The technological intentionalities discussed so far have chiefly concerned individual human-technology relations. But technologies can play a role of their own on the cultural level as well. In the last part of *Technology and the Lifeworld*, "Lifeworld Shapes," Ihde sketches out a number of character traits of the technological lifeworld to illustrate this point. The first and most important "cultural intentionality" that Ihde mentions is that technology has transformed our culture into what he calls a pluriculture. Here he cites explicitly communication, information, and imaging technologies, technologies that mediate our experience (Ihde 1990, 164–67; see also 1993b, 62). These technologies have made possible an exchange between cultures to such a large extent that they have come to play a role in the everyday life of almost everyone. Thanks to the media, we are confronted with many other cultures than our own. This confrontation does more than allow us to see what goes on in another culture from a distance; it effects an exchange of cultures on a daily basis.

Ihde speaks pointedly not of multiculturalism but of pluriculturalism. By *multiculturalism* he understands the co-existence of several cultures that in principle could exist apart from each other. The term *pluriculturalism* denotes, by contrast, that several cultures simultaneously play mutually interwoven roles in our lifeworld. Pluriculturalism goes further than the fact that we eat Chinese, Italian, and French foods, and that we decorate our living rooms with African carvings and Indonesian batiks. The pluriculturalism of the contemporary lifeworld entails that it is not enough to have a single cultural interpretive framework, a single "macroperception." We have to be able to "see" in several ways at the same time; we have to have a "compound eye," in Ihde's words, the way the director of a television program has several television screens playing simultaneously in the studio directing room. The world has become a mosaic and cannot be engaged from a single cultural interpretive framework (cf. 1993a, 114–15).

Ihde points to still another important change in our culture wrought by technological development: technologies create a "decisional burden" because of the many new choices they make possible. Having children, for instance, is no longer something that simply befalls us but has become a conscious decision. For those who are eager to have children but are unable to conceive, there are a steadily increasing number of options available. Prenatal diagnosis opens the possibility of terminating pregnancies of unwanted types of fetuses, and so forth. All of these technological developments create ever more moments, as well as kinds, of choice. And we no longer have the freedom to shirk them: "The one choice I do not have is the choice not to make a choice," Ihde says, with a nod to Sartre (1990, 181).

More generally, it can be said that technological developments produce ever more contingency (Ihde 1990, 183–84). That is, it is ever less obvious that things are "for good" what they are now, because ever more things that were once taken for granted have become controllable or at least influenceable through technological developments. The increasing pluriculturalism adds to this increasing contingency. In a lifeworld that cannot be engaged from a single way of seeing, interpretations are no longer self-evident; there are always several ways of seeing, without any one being a priori more adequate than the others.

Neutrality or Substantivism?

What is the significance of all this for understanding the relation between technology and culture? Ihde's arguments concerning the coming about of a pluricultural lifeworld, and the increasing contingency and decisional burden, weaken the instrumentalist conception of technology. Technology, after all, cannot be neutral if it is able to change a culture drastically. But doesn't Ihde thereby return us to a substantivist conception of technology as a relatively independent force? Ihde raises this question himself—"At the end of this second program, then, it appears that a different form of technological-cultural determination has reappeared" (1990, 161)—but he does not explore further the implications of this. How is his claim that there is a "different form of technological-cultural determination" related to his claim that technology is precisely not deterministic because it always has a cultural index? Does the cultural context have the last word, because a technology is what it is only inside that cultural context—or do technologies have the last word, because they have specific intentionalities by virtue of which they can change culture? To put it another way: Is the cultural relation to technologies multistable, or do technologies have a culture-changing power?

In seeking to resolve the tension between these two claims, it is helpful to reflect further on the cultural role of imaging technologies. If we try to understand the multistable cultural embeddedness of television, for instance, it is possible to say, analogously to the case of the sardine cans interpreted as ornaments for headgear, that the artifact that we call a "television" can indeed have several roles; as an occasional table, for instance, or as a display case for family pictures. But these other "stabilities" of the television-artifact do not possess the specific ability of the television: to reproduce images that were made elsewhere. Although a television can be used as a table, as soon as it is used as a television one of its "intentionalities" is to bring us in contact with other cultures, and so to contribute to the realization of a pluricultural framework of interpretation. The television thus owes its identity not only to itself, but to its context as well; but whenever it receives an identity in its multistable use context—whenever human beings relate to the artifact as a television and not as a display case—it contains its own script within that relation.

It does not therefore follow from the idea that cultural relations to the world are multistable that technology is unable to effect cultural change. Technologies are eminently able to do so, even though their role always depends on the specific context in which they function. Ihde's idea that technology can be understood only as interwoven with culture does not imply that technology cannot influence the context in which it plays a role. Technologies can do so precisely because they are always interwoven with culture. The tension between multistability and substantivism is thus only apparent. Once a relation to a technology is taken on, the relation to the technology is stable rather than multistable and the technology is able to influence the relation taken toward it, without its influence on the relation being deterministic. In principle, several cultural relations are always possible toward an artifact. But once a relation with an artifact is taken on, a "technological intentionality" arises within that relation.

4. SCIENCE AS TECHNOLOGICAL HERMENEUTICS

In *Instrumental Realism, Expanding Hermeneutics,* and elsewhere, Ihde has examined the implications of his philosophy of technology for the philosophy of science. Here, too, perception—both microperception and macroperception—plays a central role: Ihde investigates the consequences for the philosophy of science of his analyses of the role of technology in human perception.

One obvious connection between Ihde's approach to technology and the philosophy of science arises from his understanding of macroperception. Ever since Kuhn, the philosophy of science has developed in such a way as to take ever more seriously the context dependence of scientific knowledge. In place of seeking possible ways to ground scientific knowledge in reality, to find characteristics of a language adequate to speak about reality, or to discover the conditions of possibility for scientific knowledge, contemporary philosophy of science has sought to understand how the significance of scientific statements arises from the context in which they were formulated. Following Kuhn, philosophers of science have seen its development as moving, not ever closer toward a final solution to a puzzle, but rather from one framework of interpretation to another. Science always takes place inside what Kuhn called a *paradigm;* it is the work of a community of scientists who share an interpretation of reality as well as a definition of the problems deemed to be important.

Kuhn therefore did not inquire into the ground of the certainty and possibility of scientific knowledge; he relativized this ground by historicizing it. Foucault did something similar, according to Ihde, through his concept of *episteme.* Foucault's concept is less sociologically laden than Kuhn's "paradigm," and describes the "way of knowing" of a specific period specified by the language spoken by the scientists—the "discourse" with which they engage each other (Ihde 1991, 33). In the concepts of paradigm and episteme Ihde sees parallels with his own concept of macroperception. Science has to do with the ways of seeing of scientists.

This association with "seeing," however, at the same time brings to light a lacuna in the "new" (contextualist) way of thinking about science, or at any rate one aspect of science that deserves further attention in it. Science, to be sure, is to be understood as a "way of seeing," but everything said so far has localized this to the macroperceptual level only. Besides this, however, science also has everything to do with "seeing" on a microperceptual level—with concrete sensory perceptions. According to Ihde, science must not only be related to the contexts of interpretation in which it takes place, but at the same time to the sensory perceptions of scientists. And one principal characteristic of contemporary science is that these perceptions are mediated by technologies. Right here, according to Ihde, lies the most interesting connection between his approach to the philosophy of technology and the philosophy of science. Philosophers of science have readily acknowledged that scientists "see" reality in a certain way, but have paid insufficient attention to the fact that these ways of seeing are also based on concrete, but technologically mediated, perceptual seeing.

Ihde's broad thought is therefore that the philosophy of science must complement the study of the macroperceptual aspects of science with an analysis of its microperceptual aspects—and the role technologies play in these. In this way, one might say, Ihde gives a new twist to Heidegger's conviction that technology has primacy over science. Technology has primacy not because the technological mode of thinking is presupposed in scientific thinking, but because contemporary science is helpless without technologically mediated instrumental perceptions (1991, 62–63). The mediation of scientific perceptions by technological instruments is no mere accident, but has become an essential part of scientific knowledge. To understand the context in which scientific knowledge arises, philosophers of science need to extend their attention beyond paradigms and epistemes to include also scientists' perceptions and the technologies that make scientific perceptions possible.

Scientific instruments, for Ihde, are not neutral passageways to "the world itself," as should be clear from the above discussion of his analysis of the technological mediation of perception. Instruments mediate the perceptions of scientists and transform them in the process. Many phenomena studied by scientists would be unobservable without technologies. Radio telescopes, for instance, make things "perceivable" that are not accessible to the naked eye. Computer tomographs and ultrasound scanners produce images of the human body and its structures that would otherwise be unobservable. These mediated perceptions therefore reveal entities that we would never have known about but for mediating technologies.

Technological instruments, Ihde claims, play an essential role in the generation of scientific knowledge, and studying this role is crucial to understanding contemporary science. Note that Ihde tries to understand science in terms of what scientists *do*, not just in terms of the structure, conditions of possibility, and foundations of the knowledge they produce. Attending to scientific practice does not lead him, however, into a sociological or anthropological perspective, as it does many scholars in the field of science studies. For Ihde, a turn to scientific practice does not mean a turn to the analysis of the social interaction of scientists but rather to the embodiment of science in observations and in the instruments with which these observations take place. Ihde calls his position *instrumental realism.* A philosopher of science who wants to do justice to scientific practice cannot cling to a naive realism, which believes in a one-to-one relation between what scientific knowledge makes visible and what is "really" there. The reality studied by scientists is co-constituted by the technological instruments they use.

Ihde has recently elaborated the connection between instrumental mediation and the content of scientific knowledge in the last part of *Expanding Hermeneutics.* His program there is "to show how science can do a 'hermeneutics of things' by turning them into scientific objects" (1998, 139). In other words, he asks how we are to understand the scientific way of interpreting reality "in action": how reality is "prepared" by technologically mediated interpretations so that science can be done with it. For this, the classical meaning of hermeneutics needs to be expanded. Traditionally, hermeneutics was understood to involve the interpretation of texts, as well as reflection on the process of interpretation and its conditions. Ihde, however, develops a more material conception of hermeneutics. For him, it is possible to interpret things other than texts hermeneutically, and he also discerns non-linguistic forms of interpretation, such as those offered by scientific instruments. Scientific instruments constitute what scientists observe; they "interpret" reality before humans can observe it.

Borrowing a distinction of constructivism, Ihde points to two different ways in which such a material hermeneutics can be carried out: a "strong" program and a "weak" program. In a weak program, instruments are conceived as forming an interface between science and the reality it studies, co-determining how that reality is to be interpreted. A strong program goes a step further, viewing instruments as actually constituting the objects studied by the sciences and therefore co-determining the content of scientific knowledge.

The approach that Ihde characterizes as a weak program occupies itself with bringing to light the various ways that scientific observation is mediated by technologies. With this, Ihde wants to add a material-hermeneutic perspective to the post-Kuhnian thought that science is to be understood as a "manner of seeing." For Ihde, scientific observation is *"through, with, and by means of instruments"* (1998, 159), and this observation is hermeneutical in nature not only because it forms the basis of interpretations of reality, but also because the mediated seeing provided by instruments always involves, as it were, an "interpretation" performed by the instrument. Instruments prepare reality for observation: they make scientific objects out of it by making it present in very specific ways. "[T]he instrument is already *a hermeneutic device,*" Ihde concludes (1998, 149). Following a similar path as Latour, who says that scientific objects are prepared in the laboratory in and through the production of "inscriptions" that make things scientifically analyzable, Ihde argues that laboratory instruments make things "readable." Instruments prepare phenomena in reality to function as scientific objects.

This process of making things readable by turning them into scientific objects can take place in two ways. The first is by transforming something that is invisible to the naked eye into something visible. This can happen via simple magnification, as in the case of microscopes and telescopes, but also by way of more radical means of mediation. For instance, in order to make microorganisms microscopically visible, it is necessary to stain them with aniline dye. In this way they lose their isomorphism with "naked perception"—if naked perception of such small organisms is possible at all. Technologies such as X rays, ultrasound, and MRI scans go a step further: they provide a picture of the human body (or parts thereof) based on the passage of invisible rays, the bouncing of sound waves, or nuclear resonance, phenomena that are not perceivable without technologies and that thus need to be "translated" into the visible. Science is ever more occupied with things whose scale is beyond the reach of human perception, both in the microscopic and macroscopic directions: electron microscopes and radio telescopes make formerly invisible worlds visible. Ever more things that would be invisible without technology have become the objects of scientific research.

A second way in which instruments make things readable is through "text-like visualization." Text-like visualization provides a representation of reality to be "read": graphs, tables, maps, and so forth. This sort of visualization can retain an analogy with direct perception, like the mercury level in a thermometer that is high at high temperatures and low at low temperatures. But such an analogy need not take place. For instance, the way a spectrogram provides an image of a star has no analogy to the star itself.

Ihde's strong program has a more radical goal than the weak program. It not only aims to show that science always has a hermeneutical dimension insofar as it uses instruments, but at the same time that this hermeneutical dimension is constitutive of the content of scientific knowledge. Ihde indeed develops his instrumental realism as a material-hermeneutic counterpart of (social) constructivism. His principal metaphor in this connection is the "giving of a voice" to things (1998, 172). Scientific instruments "give a voice" to entities, so that they can be heard. The technology used co-constitutes the object that is investigated. Ihde calls this *technoconstitution*.

Ihde provides many examples to make clear the validity and necessity of a strong program. One is the manner in which the investigative domain of astronomy was enormously expanded by the arrival of radio telescopes and techniques to make visible invisible forms of light, such as infrared and ultraviolet. What was not directly observable became constituted as an observable object by translation technologies, allowing new phenomena to be revealed to scientific research. Another variant of technoconstitution is the use of multiple instruments to observe the same object, as the application of X ray, ultrasound, and MRI devices to study the human brain. Ihde speaks of these practices as a material variant of the Husserlian method of "phenomenological variation." Husserl's method for being able to view the "essence" of a phenomenon was to mentally imagine all sorts of variations of the phenomenon so as to be able to intuit what they have in common. Without claiming that essences can be laid bare in science—the idea of "essential intuition" stands squarely opposed to the radical hermeneutical perspective that Ihde wants to carry forward—Ihde says that science often carries out "instrumental phenomenological variations." A phenomenon is perceived throughout manifold ways with the help of different technologies, with the whole of these variations providing a picture of the phenomenon in question. These pictures do not need to converge: different instruments can offer different perceptions of the phenomenon observed.

Science originates in perceptions—but what is perceived is first prepared and made readable by instruments. Scientific observations are technologically constituted, and are not simply depictions of nature. Neglecting this technological constitution would lead to a new variant of the naive realism that constructivists warn about: the assumption of a correspondence between a scientific theory or observation on the one hand, and "reality-in-itself" on the other (Ihde 1998, 178). This assumption fails to recognize the active constituting role played by scientists through their interpretations, practices, and instruments. Until now, this role has been underplayed in science studies due to the tendency to relate the content of scientific knowledge solely to the context of interpretation in which it arose. In so doing, science studies encounters the danger of running into the opposite pitfall of naive realism: naive idealism, or the view that the ultimate font of knowledge is not the world but our ideas about it. Scientific knowledge is a product not only of interpretations, but also of the material conditions on which these are formulated—the instruments with which scientific observations are carried out.

CONCLUSION

Ihde's work offers an entirely different perspective on technology than that of traditional phenomenology. For however much Ihde and Heidegger pose the same question concerning this issue—"What does technology mean for our relation with the world?"—they arrive at completely different conclusions. Whereas Heidegger sees technology mainly as a controlling way of world disclosure, Ihde articulates a much more ambivalent picture.

The difference between Heidegger and Ihde stems from a difference in the ways in which each conceptualizes technology. As elaborated at the beginning of this chapter, the traditional project of the philosophy of technology consisted of research into the interpretative relation to reality that lies behind technology. Technology was understood as the product of a way of disclosing meaning, and this way of disclosing was extrapolated to culture in general: technology was said to imply a one-sided manipulative relation to the world. Ihde's approach to technology, however, does not begin with this world-interpretation, but with our dealings with the concrete technological artifacts, and the praxes and interpretations that are made possible by them. When the question of meaning is posed from this perspective, an entirely different picture of technology emerges.

On an experiential level, as something that mediates our experience, technology no longer appears to entail necessarily a reduction of the ways in which the world is revealed to us. There are many possible forms of technological mediation that transform our access to the world in different ways; some of these open up to us new ways of access, while others narrow this access. On the cultural level, technology is no longer seen as fostering a coordinated, uniform framework of interpretation in which the world is coerced to appear as *Bestand,* standing reserve. "[T]he predictions of analytic uniformity (Marcuse), of the victory of technique (Ellul), and even of the sheer world of calculative thought (Heidegger) are wrong. There will be diversity, even enhanced diversity, within the ensemble of technologies and their multiple ambiguities, in the near future" (Ihde 1990, 159). Technological culture does not develop in the direction of one-dimensionality, calculativity, and uniformity, but rather in the direction of plurality. Technology does not create one single way of disclosing reality—the "technological way of revealing"—rather, it fosters the proliferation of different ways of seeing within our culture. That is not to deny that important aspects of reality remain invisible within an exclusively technological way of thinking, only to say that it is a gross misunderstanding to claim that technological culture holds our entire culture in its clutches.

Moreover, technology plays a constitutive role in the production of scientific knowledge thanks to its role in scientific perception. The role of scientific instruments in the production of scientific knowledge runs far deeper than that of "depicting reality": instruments co-constitute the reality investigated. They make visible aspects of reality that otherwise would be invisible, aspects that have to be actively represented and thus "interpreted" by the instrument. In the course of elucidating this process of technoconstitution, Ihde goes much further than Heidegger did in pointing out that science must be seen as applied technology rather than the other way around.

Ihde does more justice to technology than the tradition from which he comes. Phenomenology after Husserl has always seen its main task as understanding the world in its everydayness. With respect to technology, it did not succeed in this. The role of technology in human everyday life involves far more than calculative thinking, and Ihde has begun to forge the tools necessary to understand the richness of our technological lifeworld.

NOTES

1. Ihde uses the phrase *technological* (or sometimes *instrumental*) *intentionality* in another sense as well in his discussion of mediated technologies to refer to the directionality or scope of mediated technologies. A cassette recorder, for instance, has an "intentionality" with respect to sound—a completely different one than human listeners, because it records equally foreground and background noises. Another phrase Ihde uses for this is *technological telos* (1979 77–78; 1983, 56; 1990, 102–103).

2. Phenomenologically speaking, it would be more adequate to localize "technological intentionality" not in technologies themselves but in the relation between humans and world that is mediated by technologies. By mediating this relation, technologies co-shape "intentionality"; they facilitate specific relations between humans and world, and in so doing play a role in the constitution of both. A car, for instance, is not a neutral means of going from point A to point B. It asks for specific ways of dealing with itself and with the world, and thereby helps to constitute in a specific way both the human beings driving it and the environment through which they drive. Mediating technologies are foci around which the interweaving of humans and world gets its shape, thus turning "intentionality" into "technological intentionality."

BIBLIOGRAPHY

Borgmann, Albert. 1984. *Technology and the Character of Contemporary Life: A Philosophical Inquiry.* Chicago: University of Chicago Press.

Heidegger, Martin. 1977. *The Question Concerning Technology and Other Essays.* Translated by William Lovitt. New York: Garland.

———. 1996. *Being and Time.* Translated by Joan Stambaugh. Albany: State University of New York Press.

Ihde, Don. 1979. *Technics and Praxis.* Dordrecht: Reidel.

———. 1983. *Existential Technics.* Albany: State University of New York Press.

———. 1986. *Consequences of Phenomenology.* Albany: State University of New York Press.

———. 1990. *Technology and the Lifeworld: From Garden to Earth.* Bloomington: Indiana University Press.

———. 1991. *Instrumental Realism: The Interface Between Philosophy of Science and Philosophy of Technology.* Bloomington: Indiana University Press.

———. 1993a. *Philosophy of Technology: An Introduction.* New York: Paragon House.

———. 1993b. *Postphenomenology: Essays in the Postmodern Context.* Evanston, Ill.: Northwestern University Press.

———. 1998. *Expanding Hermeneutics: Visualism in Science.* Evanston, Ill.: Northwestern University Press.

Merleau-Ponty, Maurice. 1962. *Phenomenology of Perception.* Translated by Colin J. Smith. London: Routledge.

6. Langdon Winner: Technology as a Shadow Constitution

Martijntje Smits

In 1978 Langdon Winner took a short vacation to San Luis Obispo, in central California, where he grew up during the 1950s. On a sunny December day he joined a tourist bus excursion to a nuclear power reactor under construction in Diablo Canyon. The bus followed a windy road that surmounted several hills, and when the bus reached the top of the last hill, Winner caught his first sight of the gigantic power plant with its two gleaming white cooling towers. In the background lay the capricious coastline of the Pacific Ocean. Right at that moment a huge California gray whale surfaced in the distant waves, blew out a stream of water vapor from its blowhole, and then disappeared again beneath the waves.

In the essay "The Whale and the Reactor," which is included in a collection of essays by the same name, Winner describes how overwhelmed he was by the contrast between these two powerful symbols in such close proximity—one of the power of nature, the other of the power of human artifice (Winner 1986, 168). The experience, which he describes as an "epiphany," made him realize that his fascination with the moral and political dilemmas of modern technology lay closer to his own personal and intellectual roots than he had thought. Until then he believed his political engagement to have begun during his student years at the University of California in the 1960s. This engagement continued during his summer as a "systems analyst" at the Pentagon, where he was struck by the sharp contrast between the impressive rational planning of the national defense system and the bloody, senseless, and escalating war in Vietnam that that very system was enacting. These events, he had once thought, were what led him to resolve to attend graduate school to study the meaning of the technocratic tendency in modern thought.

But the experience of the spouting whale and the reactor, he wrote, vividly recalled to his mind much earlier and more personal sources for his fascination with technocratic thinking. He recalled thinking as a child about the far-reaching changes in his environment and in the patterns of everyday life brought about by the introduction of new technologies, which had gone hand in hand with a boundless faith in new technologies. "We were technological somnambulists wandering through an extended dream," Winner writes (169). His family, like everyone else's, welcomed the arrival of superhighways, super-markets, television, prefab houses, plastics, and all sorts of household electrical appliances. He witnessed firsthand as a boy growing up in his small and rural but quickly modernizing hometown of San Luis Obispo the introduction of DDT sprayers, disinfectant soap with hexachlorophene, and X-ray machines in the shoe stores that encouraged customers to buy in exchange for the thrill of admiring the skeletons of their own feet. In retrospect, he realized that his naive and enthusiastic family members had exposed themselves to dangerous doses of chemicals and radiation.

At the time, Winner writes, he often felt uncomfortable with the uncritical acceptance accorded all the latest devices and technologies. As a child he wondered more than once whether the new technologies heralded as improvements were really such an unmixed blessing. Was the construction of a four-lane highway, which plowed straight through the landscape—sparing nothing in its path as it demolished houses and neighborhoods, farmlands and forests—really such an indisputable sign of progress? Was the milk from the industrialized dairy farm really tastier than that of the local dairy farmer?

The experience with the California gray whale and the nuclear reactor gave Winner a penetrating insight that, he writes, "said infinitely more than all of the analyses and findings of all the detailed studies I had been reading ever could" (175). The endless calculations and recalculations of the risks involved in operating the reactor and of disposing of its wastes, and the endless economic and social cost-benefit analyses of nuclear power, seemed only to obscure the real issues. The spouting of the whale made it painfully clear to Winner that the nuclear reactor is "a permanent insult to its natural and cultural surroundings"; it is "simply a hideous mistake" (176).

The message as well as the rather classical and even clichéd narrative structure of "The Whale and the Reactor"—the truth about the reactor coming not as the outcome of a process of reasoning but in a revelation contained in an instantaneous flash of deep insight, an "epiphany"—could easily leave the impression that Winner repudiates modern technology in the style of a shallow, romantic thinker. But elsewhere in his writings Winner is often critical of the romantic picture of technology and strives to undermine the simpleminded repudiation of technology in order to offer a balanced evaluation of it.

The informal and affecting style of this essay is characteristic Winner. The choice of a nuclear reactor is also typical. The whale is new, but this is not the first time that a nuclear power plant surfaces in his work.[1] It is no accident that the nuclear plant enters his personal story in 1978. At this time nuclear energy was the most popular target of the alternative political movements and critical social theorists in whose circle Winner traveled. Nuclear energy posed, to be sure, a serious environmental question, but this fact does little to explain the extensive, powerful, and sometimes violent protests against it, given the existence of other extremely pressing environmental issues. The protests at that time were attracted by the great symbolic significance of nuclear energy, and the way it served as a metaphor for much of what was wrong with society and modern technology.[2] Nuclear energy stood for centralization, technological complexity, danger, and for all of the tendencies that environmental activists and critics of technology distrusted the most: the increasing power of experts, the threat to individual freedom, and the strengthening of totalitarian power.[3]

The technology of nuclear power is a classic example of Winner's central thesis that "technical things have political qualities" (19). He defines politics as "arrangements of power and authority in human associations as well as the activities that take place within those arrangements" (22). By technology he means not technological knowledge so much as tangible artifacts, the products of technological processes (22). What holds for nuclear energy applies as well to technology in general; it has political qualities or properties in the sense that it brings about a reconstruction of social roles and relations and forces us into specific power relationships. Technological objects and systems form what I might call a "shadow constitution," a hidden political power in society, an unwritten set of laws that establish social roles and relations.

But there is a second sense in which technology can be said to have political properties. Technology not only establishes specific social roles and relations, but also has a more general influence on everyday "forms of life" (11). Technologies often affect not only everyday behavior but also norms and values, our self-understanding, our perception, and our ideas about space and time. The impact of the technological grip on forms of life is evident not only in things like automobiles and televisions, which are now all but indispensable ingredients of everyday existence and perception, but also in new medical technologies, which have changed not only medical practice but also the very definition of medical care and human health.

Winner uses the case of nuclear energy to illustrate a second thesis: namely, that the political implications of many current technological systems strongly conflict with prevailing liberal-democratic ideals. According to Winner, for instance, the management of the risks of nuclear energy, especially when plutonium is used as a fuel, is accompanied by a sacrifice of civil liberties, for extraordinary safety measures must be taken to protect against terrorism (37). Moreover, the use of nuclear energy implies a centralization and consolidation of the power of experts, depriving citizens of their ability to exercise influence and control over the social good.

The issue of nuclear power plants will lead us into the core of his critique of technology.

THE POLITICAL DIMENSION OF TECHNOLOGY

Since 1985, Winner has been a professor at the Rensselaer Institute in New York, a technical university, where he teaches political science in the department of science and technology studies. Before that he taught at the Massachusetts Institute of Technology, and he has held guest professorships in Leiden, The Netherlands (1973–74) and Oslo, Norway (1991–92). From 1991 to 1993 he was the president of the Society for Philosophy and Technology. Winner is also known as a connoisseur and critic of modern music. He was an artistic advisor of the minimal music film *Koyaanisqatsi*, which depicts the alienating effects of mass culture. We can therefore presume to recognize Winner's hand in the scene in which behind an idyllic beach looms a gleaming nuclear reactor.

Winner has addressed himself to the political dimensions of technology ever since the beginning of his career as a writer. His work is not extensive: it consists of one book, *Autonomous Technology* (1977), one collection of his own essays, *The Whale and the Reactor* (1986), which includes his most famous article, "Do Artifacts Have Politics?" and a number of short articles. According to his site on the Internet, a third book is in preparation on "the politics of design in the contexts of engineering, architecture, and political theory." Winner has also edited a volume of essays entitled *Democracy in a Technological Society* (1992).

The issue that threads through all of Winner's work is the meaning and influence of modern technology on the organization of human social life. This issue raises the question of the relation between modern technology and the key tenets of liberal-democratic politics as it is currently practiced. This question is already prominent in his first book, *Autonomous Technology*, in which his political perspective on technology appears in the form of his response to a number of prevailing, critical views of technology.

First among them is the conventional instrumental perspective, in which technologies are conceived as a neutral means to help carry out a specific practice, while denying that they frequently transform this practice in radical ways. "Technology assessment," an approach to evaluating new technologies, belongs to this perspective. The approaches of this perspective ask in which direction technology must be steered, taking into account "side effects" such as risks to health and the environment, and with economic costs as the most decisive issue. The "side effects," as the phrase indicates, are treated as secondary and as not necessarily connected with the functioning of the technology in question. They appear to be removable, with the technology itself able to remain intact. The undesirable effects can be kept in check by, for instance, "the opportunity to mount campaigns to modify or block the planning of ill-considered urban highways, to regulate the manufacture and distribution of dangerous insecticides, or to alert the populace to the dangers of propellants in aerosol cans." However, these approaches to technology "treat only symptoms, but leave the roots untouched" (1977, 107). According to Winner there is ultimately nothing secondary about the phenomenon of side effects in technology, the changes it brings to the texture of modern life. "It is, in fact, the most important accomplishment of any new technology" (1986, 11).

But Winner is equally hostile to many of the philosophical critiques of technology that offer a "depth analysis" (1977, 108) of the crisis of technological culture, or approaches that seek to discover a "new ethics" for the technological era. Winner also rejects the view of technology as a single, monolithic phenomenon à la Ellul, Husserl, Heidegger, and Scheler. As soon as they have plumbed the depths of the essence of technology, their speculations tend to make "any question of the sort 'What next?' largely irrelevant" (1977, 131). Moreover, these thinkers wind up concluding that human beings have no ability to act in ways that will alter the course of technology. And because of the inherent reductionism involved, Winner is also skeptical of Lewis Mumford's analysis in which the origin of the present-day unchecked growth and dominance of technology is to be sought in the "lethal link" of, on the one hand, the innate greed and lust for power of human beings, and on the other hand, their modern self-image, which tells them that they are essentially tool-making creatures.

Thinkers who trace the dominance of technology or technological rationality back to specific cultural roots in the Western tradition (be they Christian, Greek, or other), such as John Passmore, Lynn White, and Max Horkheimer, leave themselves open to the charge of reductionism. They fail to do justice to the complex, multifaceted history of technoculture, and they usually end by calling for a new ethics or a new fundamental change in attitude, with

White a good example. But "the great mistake of attempts to propose a new ethic . . . is that they proceed as if the enterprise were something like philosophical engineering—a special kind of problem solving. But at the level at which one seeks to elucidate first principles about man, nature, and being, the world is not a problem set. It is, if anything, a question set" (1977, 133).

In short, Winner thinks, we must avoid the "depths without direction" offered by many philosophies of technology, as well as the "details without meaning" provided by instrumental perspectives (1977, 134). In seeking to describe the meaning of modern technology for human social life, Winner turns instead to political theory, "a subject that has traditionally found its work in the middle ground between ideas sublime and affairs quotidian" (1977, 134).

This is not to say that one can find in existing political theory an adequate, ready-made critical evaluation of technological culture. On the contrary, even though political philosophy has concerned itself since the beginning with the question of the just society, it has also refrained from the outset from reflecting on the influence of technology on the quality of that society. Winner describes how both classical and modern political philosophy have kept technology and politics apart by placing them in separate domains (Winner 1992, 343).[4] While political philosophies have evaluated technology and politics very differently over the years, depending on what these philosophies see as the best way to realize human freedom, they have still always held them more or less apart.

In classical philosophy this contrast is clearest in Aristotle, who set the technical activity of the practical arts (poiesis) apart from the political, public activity of action and speech (praxis). Aristotle elevated political activity above technical activity because the end of life was happiness as the outcome of self-realization in a moral community. This could be achieved only by intensive participation in political life. While politics was intimately bound up with the realization of human freedom, technology was associated with material necessity and with the eternal struggle against natural forces.

In modern political philosophy, on the other hand, the evaluation of technology as the making of things, and politics as public action and speech is radically inverted. The prestige of technical activities has sharply increased in the wake of the belief in progress through reason, while politics in its traditional meaning has fallen into discredit because it has come to be seen as the kind of thing that causes wars and social disruption. Politics is perceived as needing to remodel itself along the lines of technical activity, as the means for bringing about certain and predictable results in the social sphere. The trust in technical activity and its elevation over political activity is heralded by a remark of Francis Bacon comparing the contributions of politicians and scientists to human history: "The benefits of discoveries may extend to the whole race of man, civil benefits only to particular places; the latter not beyond a few ages, the former through all time."[5] Freedom and power are no longer associated with politics, but with goal-directed rationality. The progress of technology will liberate humanity from physical and social burdens.

Although twentieth-century philosophers of technology have in several respects moved beyond the assumptions of modern philosophy, the firm separation between the technical and political domain remains alive and well for many of them—and those who claim that it does not exist anymore lament its passing. This is the reaction of political philosopher Hannah Arendt, for instance—whom Winner frequently cites, though notably enough not in this connection—who describes the loss of the political, public sphere in contemporary mass society as a consequence of modern technology and its associated consumer ideology, which elevates technical production over political action (Arendt 1958). In place of that she pleads for a rehabilitation of political action in the Aristotelian sense, which therefore amounts to a straightforward reversal of the new scheme and a restoration of the old.

Winner's political perspective on technology allows him an almost dialectical perspective on the leapfrogging of "technology" and "politics," for his view that technology has political properties allows him to see them as blending into each other.

Winner, of course, is not the only one to have thematized the political character of technology. Indeed, an entire academic field, usually called "Science and Technology Studies," or STS, has applied itself to pointing out and elaborating the many ways in which technology is not politically and morally neutral. STS has become a catch-all name for a diverse collection of sociologists, anthropologists, and historians who, in contrast to the majority of philosophers of technology, study the dynamics of the development of science and technology in empirical ways. Its devotees have been able to show how the capricious, complex forms of modern technology are the outcomes of decisions in which social and political factors play an important role. Winner is a fellow traveler in this company, largely because he was one of the first to highlight the political ladenness of technology, but at the same time he has never fully identified himself with it. In contrast to most STS researchers, Winner has remained strongly oriented by traditional political theory and the philosophy of technology. His critical distance from STS and in particular from the social constructivist current within it appears most strongly in his controversial article entitled "Upon Opening the Black Box and Finding It Empty," to be discussed below.

The philosophers who were characterized in the introduction to this book as "classical philosophers of technology," such as Heidegger, Jonas, and Arendt, but also Ellul and Mumford, do have one essential correspondence. Though they differ in many ways, they each see a sharp opposition between technology and the technological approach and another human capacity, which they describe in different ways. They all agree that, most essentially, human beings are not *homo faber* (instrument-makers) but rather symbolic, linguistic, or political beings. Furthermore, these features are the principal condition of culture, though this culture is threatened by the growing technologization of the world. This opposition between technology and culture forms the basis of their critique of technology. A characteristic of modern technology is that it is insensitive to the human condition and threatens to level it. The above-mentioned thinkers might well be called "humanistic philosophers of technology," because the human condition as they see it functions not only as a description but also as a humanistic norm or value. They plead for a new kind of technological development that serves these humanistic values.

Winner's position that modern technology has political properties takes an important step in the direction of moving beyond the dichotomy of technology and culture. The remainder of this essay will focus on this promising development, and on the question of whether Winner has indeed succeeded in freeing himself from key assumptions of the classical humanistic philosophy of technology. Has he in fact achieved a detached evaluation of technology and technological rationality and done justice to the historical complexity and ambivalence of technological developments, without at the same time giving up the philosophical ambition to provide a critique of modern technology?

In the paragraphs that follow I shall explore the most important respects in which Winner's vision of technology sets itself apart from that of the classical philosophers of technology, in order to clarify the status of his work in contemporary philosophy of technology.

A LOVE-HATE RELATIONSHIP WITH ELLUL

We can learn the most about Winner's relation to the classical philosophers of technology from *Autonomous Technology*, in which he engages practically all of them in discussion. Winner sees Ellul, of whom he is clearly a critical admirer, as having a central place among them. Thus the following discussion will focus on Ellul and the differences between Ellul's work and Winner's.

Autonomous Technology is a comprehensive and fairly wide-ranging study of the different forms of the notion of the autonomy of technology and technological development. This notion implies that technology follows its own laws and does not allow itself to be governed, or to be governed only, by human beings. Technology is "out of control." This conception thus amounts to an almost complete reversal of the familiar instrumental view of technology as slavishly obeying the will of its human masters in order to increase their freedom and power over nature.

The conception of technology as possessing an autonomous power had been articulated as early as the beginning of the nineteenth century by philosophers and writers of the romantic tradition, who had noted the glaring contrast between the prevailing view of machines as benign tools and their real impact on daily life. Machines frequently not only refuse to be subservient; they even often generate and enforce norms for the conduct of their human "masters." This situation has certain parallels to the famous master-slave relation: while human beings seek to use technology to free themselves from their dependence on economic necessity, it is paradoxically technology that turns around and makes them dependent, taking away their freedom once again.

According to Winner, the image of "autonomous technology" has been drawn in its most uncompromising and influential form by Ellul. Ellul, who is known for his fatalistic and deterministic vision of technology, maintains that modern technology has become autonomous in the sense that "it has fashioned an omnivorous world which obeys its own laws and which has renounced all tradition" (Ellul 1964, 14). Technology has created a completely new culture, in which traditional norms and values have been abandoned. The goals of technological society have become abstract and are no longer questioned. All traditional values are subordinated to a single one, the instrumental value of efficiency. Technological development thereby acquires a focus and an internal dynamic that is independent of contingent human decisions. Attempts to escape the grip of technology are doomed to failure, because even these are accessory modes of technology. Although Ellul does not deny human freedom, his position is that it cannot be realized in the shadow of the technological dynamic. Ellul belongs in the ranks of the humanistic philosophers of technology due to his belief that human beings shoulder the responsibility for trying to regain the freedom that technology has swallowed up. Ellul relies here on a Christian-inspired vision, but provides little in the way of content for how to achieve it.

Throughout the book, Winner's feelings toward Ellul oscillate back and forth between love and hate. Winner does not try to conceal his admiration for what he calls Ellul's "challenging metaphors," and calls *The Technological Society* "a fascinating, sprawling masterwork in the literature of autonomous technology" (1977, 41). As the starting point of his own book, Winner adopts Ellul's insight that technology is the central phenomenon of the modern era, in which the local character of traditional technologies, which take on different forms in different circumstances, has been completely swallowed up. But at the same time Winner is equally insistent that Ellul's position is unrealistically extreme due to his deterministic and reductionist conception of technology, or technique, as Ellul calls it. In Ellul's definition, technique is "the *totality of methods rationally arrived at and having absolute efficiency* (for a given stage of development) in *every* field of human activity" (Ellul 1964, xxv; Winner 1977, 9). In this conception, every social phenomenon—even democracy and the formation of public consensus—is interpreted as a product of technology.

Nevertheless, Winner's embrace of Ellul's starting point is qualified, with points of difference scattered here and there throughout the book, while Winner's repudiation of Ellul's extreme vision is modulated. For example, Winner challenges Ellul's tendency to portray technological development as following a linear path, referring to the results of studies in the history of science and technology. Winner also disagrees with Ellul's claim that there is no longer any difference between modern science and technology, or between the invention and application of technology. Winner does not even share Ellul's conception of the cause of the triumph of technology. It is true that, in contrast with the philosophers of technology mentioned above, Ellul does not attempt to trace this triumph back to a human deficiency or to a specific cultural root of modern civilization, attempts that Winner would reject as inherently reductionist. Ellul argues on the contrary that the conjunction of several different fairly recent historical circumstances lead to the crystallization of the modern technological world view. According to Winner, however, in the end Ellul still falls into the trap of reductionism, thanks to a fairly superficial interpretation of history in which all historical tendencies point in a single, unequivocal direction: the triumph of technique.

But Winner finds these points of difference to be of minor importance to Ellul's overall significance. It is all too easy to fault Ellul for his slipups, as his critics enjoy doing. Here and there Ellul's conceptual leaps overreach—but it is precisely because of his theoretical rigor that he was able to arrive at his powerful vision of the development of technology as a totality.

Winner finds that a more important point of difference between himself and Ellul lies in the latter's conception that the development of modern technology is headed toward the integration of technological systems into a unity that will be driven by technological rationality and where the power will be concentrated in the hands of a small group of experts. This concentration of power, according to Ellul, is a necessary consequence of the need for planning, coordination, and integration dictated by technique.

Ellul's ideas on this point closely resemble those of a number of thinkers of whom Winner is critical, including Price, Galbraith, and Marcuse, who were leading spokesmen in the technocracy debate that raged during the period in which Winner wrote *Autonomous Technology*. This debate turned on the claim that modern society must be understood as a *technocracy*, in which political power is concentrated in elites of technological experts, who no longer arrive at decisions through the processes of political negotiations but rather through the application of scientific rationality. Winner completely rejects this picture. The proponents of technocracy, he claims, pose the wrong questions: not "Who governs?" but rather "What governs?" is the right one to ask. What they overlook is that power in the technological arena is not exercised by individuals or elite groups but by technological systems. Those technological powers are implicitly but coercively deciding the important political issues shaping moral and political life.

In contrast, Ellul asks the right questions, Winner says, even if he doesn't arrive at the right answers. Against both Ellul and the proponents of technocracy, Winner sees power in modern society as being "fragmented" rather than centralized, spread around the different systems that nest in the technological order. What governs is not Ellul's technique—monolithic Technology—but many technologies. Technologies rule the state, rather than the other way around.

Winner calls this condition—in which political life is seen as taken over by technological systems rather than technique—"technological politics." The political influence of modern technologies is strengthened by two mechanisms. The first is *technological imperative,* or the tendency of human needs to translate into the need for more technology—for instance, more electrical supplies or · energy resources. The second mechanism is *reverse adaptation,* or the way the original goals of an activity can adapt to and become modified by the technologies available for carrying them out. One of Winner's favorite examples concerns the race to the moon. Ever since the moon landings, NASA has continually proposed new projects, including the space shuttle and space missions to Mars, Venus, and elsewhere in the solar system, and in defending these projects has invoked the argument that they are needed in order not to have to dismantle the existing technological and industrial capability to make spacecraft. Thus the original means—spaceflight technology—has become an end itself, for which additional ends must be sought in order to justify continued maintenance of it as a means.

In Winner's discussion of *technological politics* another important difference emerges between his work and Ellul's, one that Winner does not elaborate. In contrast to Winner, Ellul does not describe technology in terms of political power, meaning that Winner's evaluation extends into dimensions that Ellul's does not. The autonomous power that Ellul attributes to technology is an apolitical power that does not appear to have human origin and over which humans appear to have little control. But for Winner the power of technology is through and through political; it is always the result of human choices. Accordingly, if we want to change technology, it must be by political means.

RAISING FRANKENSTEIN

Winner wants to avoid leaving the impression that he is a technological determinist in the way of Ellul. Determinism is far too drastic and fails to do justice to the choices and actions that are really open to us to escape from the grip of technology. Rather than viewing human beings as in the clutches of an all-embracing "Technology," Winner describes the grip technology has on the lifeworld through his notion of human beings as "technological somnambulists" (1986, 10, 169). This term echoes Ellul's call for the sleepers to awaken at the beginning of *The Technological Society*. The metaphor is ill-chosen, however, for a sleepwalker is disengaged from consciousness and will, while Winner speaks explicitly of willful acts—"we willingly sleepwalk" (10)—rather than of passively suffering our technological fate. Another, still more convincing analogy is the metaphor of *forgetfulness*, which Winner uses in the same book to apply to the failure to recognize the conditions under which technology functions.

Winner illustrates his interpretation of the apparently autonomous grip of technology with the aid of Mary Shelley's novel *Frankenstein* (*not* the story as presented in the film versions). According to Winner, the lesson that Shelley wants to present is that, if not carefully and completely supervised, technological products can lead to horrible outcomes. A key passage in the book is the moment when Victor Frankenstein, the monster's maker, first realizes that his invention has indeed come to life, and he flees in fear. From that moment on, Frankenstein invests all of his energy in forgetting his invention, instead of taking up his responsibility for his own creation and treating it as a small, innocent child, to be raised and civilized as a participating member of society. The terrible deeds that the monster then goes on to commit in the novel are ultimately due to the escapism of Frankenstein himself, who time and again shirks his duty to supervise his creation.

In invoking the monster Frankenstein, Winner does not mean to suggest that technology is inherently monstrous; the monster became what he was not by nature but by nurture. Rather, Winner's attention is focused on his maker, Victor Frankenstein, whose situation is our own, "for there is a sense in which all technical activity contains an inherent tendency towards forgetfulness" (Winner 1977, 315–16). This nonchalance is fostered by the immediate availability that we demand of technology. The moment a technology becomes available, we tend to stop worrying about its development or about the details of its operation and want it to be simply useful. We no longer consider it necessary to understand the conditions of its functioning—or if we do, we forget about it as quickly as possible. "This more than anything else," Winner says, "is the true source of the colossal passivity in man's dealings with technical means" (315).[6]

What kind of careful supervision of technology would Winner find acceptable? Around the time Winner was writing *Autonomous Technology,* many Western countries were beginning to require risk and cost-benefit analyses of new technologies and large-scale projects as a condition for their approval, such as "environmental impact statements." This requirement does indeed exhibit some concern for the impact of technology. But Winner is deeply suspicious of such analyses, finding that they tend to grossly underestimate the implicit imperatives built into new technologies. The regulative approach, he finds, fails to recognize the political character of technology, and as a result is of only marginal value in perceiving, let alone coping with, the long-term impact on society.

But Winner is also unconvinced by analyses that claim to take into account the political character of technology, finding "overtones of utopianism and unreality" (1977, 327) in proposals for alternative, flexible technologies, for the direct participation of users in their developmental processes, and for building into them a higher degree of flexibility and mutability. The revolutionary ambitions of such proposals will inevitably be thwarted because they fail to take into account the fact that social implications are to a large extent, and invisibly, already built in to technology. Such proposals also ignore the fact that one can never confidently know in advance whether one or another technological system will have a better social impact.

Winner concludes his argument in *Autonomous Technology* with a call for a more experimental approach allowing the acquisition of insight into the hidden structures and conditions of contemporary technological systems. Prior to approval of an alternative technology, the system must be "dismantled," a method that Winner calls "epistemological Luddism." Luddites were nineteenth-century protesting laborers who destroyed new textile machines; Winner's qualification, "epistemological," indicates that he means not a literal smashing but rather a deconstruction of the social and political conditions of technological systems. According to Winner, this can be done first of all by conducting experiments in what happens when we do without different ways of using technologies like automobiles, television, and telephones. This would allow people to investigate and reflect on the ways in which technologies shape the relations between human beings and the needs and dependencies technologies create. A second way of dismantling technological systems is by refusing to fix them when they fail, in place of the usual method of patching them up with more technology, in accordance with the technological imperative.

Unfortunately, aside from a few fleeting remarks, Winner fails to provide any descriptions of the kind of experiments he has in mind. Aside from the question of whether his own remarks aren't at least as utopian as the movement for alternative technologies, the remarks also reveal a remarkable counterpoint to the message of the Frankenstein story. For by counseling epistemological Luddism, Winner has in mind not educating and civilizing technology but banishing it—experimentally, at least. In yet another echo of Ellul, Winner seems to be pointing in these remarks in the direction of a technological asceticism.[7]

Remarks in his later works, however, appear to be more in harmony with the pedagogical message of the Frankenstein story. In several recent articles he calls for more public participation in technological development and for more discussion of the political ramifications of the design processes (e.g., Winner 1991, 291; Winner 1992b, 355). One of his inspirations is a Swedish plan involving the development of a new, automated system for laying out newspapers in which different groups of consumers and employees from the graphics industry were involved in the decision process from the outset (Winner 1992b). In contrast to epistemological Luddism, this kind of approach is oriented toward intervention and participation in the process of technological development, and therefore has more resemblance with raising monsters. For raising is always a continual, forward-looking process and not a retrospective critique of the circumstances in which the child has grown up, such as epistemological Luddism suggests.

THE POLITICS OF ARTIFACTS

Although the message of *Autonomous Technology* is that technology must be described as a political phenomenon, the book is short on illustrations of what the political effects of technologies might be. In this respect Winner resembles the speculative, less empirical style of the classical philosophers of technology. But this changes. In his collection of essays *The Whale and the Reactor*, Winner hones and dramatizes his political interpretation of technology with the aid of a number of incisive examples, the most famous of which concerns regional planner Robert Moses's overpasses in the famous article "Do Artifacts Have Politics"—an article that may well owe its fame to this example alone.

In the article, Winner distinguishes several ways in which technological artifacts can have political properties. A first category involves artifacts that are intentionally supplied with political effects, and the examples Winner provides include the low-hanging overpasses that Moses had built over the verdant parkways that led from New York City to the beaches of Long Island. The low hang of the bridges prevented public busses, serving mainly low-income groups, from traveling on the parkways, effectively prohibiting racial minorities and the poor from reaching the beaches, while whites from the upper classes who owned automobiles had no such problem. According to evidence collected by Moses's biographer, Moses had the bridges built that way with this social bias and racial intent explicitly in mind. Winner remarks that although the political impact—social injustice—of these quite concrete bridges was built into this particular system, it was not a necessary linked to the functioning of the bridges as such. A bridge has a certain flexibility regarding its political effect, in the sense that building them in other ways would have had a different political effect. But this seems to weaken Winner's case. For while he offers the example as evidence against the instrumental view of technology, it could just as easily be interpreted as evidence for it, inasmuch as it is an instance of a technology in the service of certain ends (in this case political) that shape its development.

A stronger challenge to the instrumental view, however, is found in the second category that Winner distinguishes, involving artifacts in which political effects are unintentionally and unavoidably built in. As an example of this Winner refers to the mechanical tomato harvester, which was developed in California at the end of the 1940s. Because existing tomatoes were too vulnerable to being bruised and crushed by the machine, agricultural researchers developed new varieties of tomato—hardier, but less tasty—to be used with it. The happy owners of the tomato industry reaped a considerable cost savings. But this economic benefit was not equally shared. Ultimately and unintentionally, the new technology brought about a radical reconstruction of the social and economic relations in the tomato industry. The large and expensive machines promoted a concentration of the tomato industry, leading to greater productivity but a steep reduction in the number of tomato growers and the elimination of tens of thousands of jobs. "What we see here instead is an ongoing social process in which scientific knowledge, technological invention, and corporate profit reinforce each other in deeply entrenched patterns, patterns that bear the unmistakable stamp of political and economic power" (Winner 1986, 27). But even the development of this tomato harvester is relatively flexible in its political effects. As in the case of the overpasses, alternate developmental paths are thinkable.

But in the third category of the politics of artifacts this is no longer true.

In it, the social and political effects are more or less intrinsically linked to the technologies themselves. The functioning of these technologies requires specific power relations, and all flexibility disappears. In contrast to the overpasses, which can function in many different political contexts, something like a large ship can function only when there is a strict hierarchical relation among those on board. Winner does not go so far as to claim that specific political relations flow necessarily out of this third type of technological artifact, but rather that the artifacts exert strong pressures in the direction of specific political structures with which they are most "compatible." The generation of solar energy is said, for instance, to fit best a decentralized, democratic structure, because the solar cells can function on a small scale without losing much in the way of efficiency. In the case of nuclear power plants the reverse is true; the larger the scale the better, to maximize efficiency. Winner cautions that control of such technologies collides with democratic ideals of freedom, which is especially true in the case of nuclear weaponry: "The internal social system of the bomb must be authoritarian; there is no other way" (Winner 1986, 34).

Winner's account of technology provoked a sharp reaction from Steve Woolgar, a prominent sociologist of technology. According to Woolgar, Winner generally presents the political consequences of technology as unambiguous and predictable, while Woolgar holds that these consequences are always the contingent and unanticipated result of a capricious social process. Winner uses the example of Moses's low-hanging overpasses to suggest that the effect of a developmental process logically flows from the ideas of the developer. But the results of much STS research suggest that the effects of a technology usually deviate from the original intention, and that one and the same technology, far from being unambiguous in its effects, can have a wide spectrum of effects (Woolgar 1991, 34). According to Woolgar, the political identity of technologies is not fixed, but has an "interpretive flexibility." The significance of artifacts cannot be established ahead of time, but is a result of social interaction.

Winner has defended himself against this charge by pointing out that the political effects of the artifacts he was describing are not as arbitrary as Woolgar claims (Winner 1993). The difference between their two positions boils down to this: For Winner, the political structures that are promoted by technology are much more coercive and inflexible than Woolgar supposes. Winner is prepared to admit that the effects of technologies cannot be established ahead of time, but also claims that he has given solid arguments in support of his contention that social injustice was a consequence of the low-hanging overpasses. Woolgar's discounting of the influence of Moses on this consequence of his overpasses Winner ascribes to "political naivete." The social constructivist notion of "interpretive flexibility" amounts, for Winner, to an unintentional manifestation and extension of the value relativism of the evil instrumental conception of technology.

THE EMPTY BOX OF SOCIAL CONSTRUCTIVISM

This response to Woolgar is part of an article entitled "Upon Opening the Black Box and Finding It Empty," in which Winner outlines his objections to social constructivism, an important current in the STS field. Winner uses the term *social constructivism* as a collective name for the work of a diverse group of thinkers that includes Harry Collins, Trevor Pinch, Wiebe Bijker, Woolgar, Michel Callon, and Bruno Latour. Despite their different emphases, the starting point and outlook of these thinkers, Winner says, is "fairly consistent" (1993, 366). What unites them is the conviction that the study of technology consists of opening technological "black boxes" through more careful investigations into the internal social dynamics of specific technological developments than have been hitherto conducted.

Winner raises the question of what contributions social constructivism can make to an "understanding of the place of technology in human affairs." He emphasizes that social constructivism has a number of valuable sides; the opening of technological "black boxes" is of unmistakable importance. This is hardly surprising, given that this approach resembles the deconstructive attitude embodied in Winner's own "epistemological Luddism," which also aims to unearth the social and political conditions of technology. By opening black boxes, social constructivists were able to refute some persistent ideas about technology development, such as that technologies are the work of "great men," or that a clear boundary can be drawn between the social-political and the technological domains. Above all, social constructivist studies have shown that there are far more alternatives and decision points in technological development than is usually assumed. As a result, Winner says, those who want to reflect on "the place of technology in human experience" cannot easily neglect these studies.

Despite his general sympathy with the aims of social constructivism, Winner finds that it has little to say about the crucial question of technology's impact on the quality of everyday life and on the quality of social relations; indeed, he claims, it harbors an "apparent disdain" for such evaluative questions. This deliberate turning of a blind eye to moral issues is, according to Winner, reinforced by the limited perspective that social constructivists have on technology: they focus only on the dynamic of the interaction of social groups in their case studies, and ignore the deeply rooted historical patterns of technological development that underlie these local processes. Those philosophers of technology such as Marx, Mumford, Heidegger, and Ellul who have reflected at length on these patterns are dismissed by the social constructivists simple-mindedly as old-fashioned. However diligently the social constructivists pursue their opening of technological black boxes, all they have to show when they do is an empty box. In contrast to the classical philosophers of technology, the social constructivists give us no foothold for evaluating the different possibilities that technology creates.

Following the Woolgar-Winner exchange, a Swedish sociologist named Mark Elam took up cudgels in defense of social constructivism and of Woolgar in particular (Elam 1994, 101–106). Elam reproached Winner for his "metaphysical position" in insisting on the importance of judging the moral value of technology. Woolgar in contrast guards against taking a firm public stance on moral and political issues, because this can only lead to violence against all those who do not share his point of view. Woolgar's method, Elam claims, far from being morally empty and politically naive takes its well-thought out point of departure from the moral and liberal stance according to which there is no definitive truth.

In replying, Winner says that he does not, as Elam claims, presume the existence of universal moral standards (Winner 1994). Moral evaluations of technology should spring from dialogues occurring in local communities around specific issues. Rather than wanting to propose general truths, Winner wants to widen the political space in which choices about technological developments are made, to include all citizens affected by the consequences of the technology. In this connection Winner has also referred to the republican ideals of Thomas Jefferson, who called for the full political participation of the citizenry (1992b, 350), even though Winner recognizes that the structures and institutions of contemporary life make it difficult to realize the full involvement of everyone in moral and political questions concerning technology.

Winner's obvious aversion to moral standards cannot conceal that he is still appealing to an unmistakable moral standard. The standard against which Winner measures technologies is that of their democratic potential. "We should try to imagine and seek to build technical regimes compatible with freedom, social justice, and other key political ends. . . . What I am suggesting is a process of technological change disciplined by the political wisdom of democracy" (1986, 55). He calls for the development of new political roles and institutions in which the spectrum of possible choices concerning technologies can be weighed by the entire community rather than by experts. Rather than being a metaphysician, Winner is a believer in direct, local democracy, and thus belongs to a movement that continues to this day especially among North American thinkers.[8]

One consequence is that Winner would entrust judgments about technology, not to philosophers of technology, but to the citizenry as a whole. This remark appears to conflict with Winner's assertion in "Upon Opening the Black Box" that philosophers of technology play an important role in fostering a moral sensibility in modern technological culture. But there is no contradiction once one recognizes the distinction Winner draws between characterizing the large patterns of technological development—which philosophers should continue to do—and the democratic process of judging specific technological developments, which should be left to the citizenry, with the aid of the framework provided by the philosophers. This distinction then raises the question of whether there can in fact be a clear difference between philosophical interpretation and democratic judgment, a question that, so far at least, Winner has not even mentioned.

TECHNOLOGY AND DEMOCRACY

Winner's response to Elam reveals some aspects of Winner's own moral vision and judgment on technologies, and the possibilities that they open up and close off. In his works up to and including *The Whale and the Reactor*, Winner has only hinted that the politics that technologies embody are often not the sort of politics that he prefers. He has not been explicit in articulating his ideas of how to handle the democratic criteria to differentiate between the desirable and undesirable politics of artifacts. On the basis of Winner's own writings, it is impossible to render a comprehensive judgment about monolithic Technology, given that each technology can embody a different politics or form of life. At the same time he claims that philosophers of technology must provide citizens with a meaningful vision of technology to help them make judgments about it. What, then, is Winner's own position?

Clues may be found in an article entitled "Artifacts/Ideas and Political Culture." In it, Winner makes it loud and clear that he thinks that the political effects of many contemporary technologies are mainly "in flagrant contrast" to the prevailing liberal-democratic ideology. According to the liberal-democratic ideal, "human beings flourish, achieving what is best in their potential, under conditions of freedom, equality, justice and self-government" (1991, 289). Most political messages that are embodied in technologies cannot be reconciled with liberal ideology, for they tell us: "Power is centralized. The few talk and the many listen. There are barriers between social classes. The world is hierarchically structured. The good things are distributed unequally. Women and men have different kinds of competence. One's life is open to continual inspection" (1991, 288).

At this point one begins to suspect that Woolgar might have been on to something after all in his complaint that Winner moves a little too hastily in his interpretations of the politics of technologies. "Artifact/Ideas and Political Culture" contains no examples, and lacks a careful estimation of the ambiguities that no doubt arise in the interpretation of the political import of technologies. Winner appears to fall back on Mumford's dualistic schema in which there are only two basic types of technologies: one authoritarian, large scale, and system centered, the other democratic, flexible, and able to be tailored to the human scale. The features of modern technology that Winner enumerates all fall under the first kind. When Winner finally comes to make a judgment about modern technologies, he neglects his former sensitivity for reductionist arguments that fail to do justice to historical complexities.

CONCLUSION

Winner's work searches to work out a middle path between the philosophy of technology on the one hand, which has carefully worked out visions of technological culture, but which is prone to fall victim to reductionist and deterministic positions, and social constructivism on the other, which is able to do justice to empirical and historical complexity but fails to evaluate the moral dimensions of technological development. Does he succeed?

One might remark first of all that Winner has performed an important service in pointing out clearly how imperative it is to find a middle path. But the weaknesses of his "Artifacts/Ideas" article also indicate how tricky it is to actually walk this middle path without winding up once again stuck in one of the embankments; it might be more difficult than Winner himself recognizes. Winner acts as though the two sharply diverging approaches can be easily combined, but does not stop to ask whether this might not be as difficult as mixing oil and water. It may well be that these two genres of technology studies are mutually exclusive. As it now stands, the methodological approach of social constructivism is not open to making global moral judgments about technological culture.

The embankment in which Winner himself tends to wind up is that of the humanistic philosophers of technology, who start with the assumption of a radical dichotomy between technology and culture. The humanistic ideal that Winner feels needs to be defended against the encroachments of technology is that of a democratic, just society in which each individual realizes his or her potential through participation in public debate. Put that nakedly, technology and democratic politics, despite Winner's original intentions, seem to wind up again in different camps. With his desire to have technological change disciplined by the political wisdom of democracy, Winner seems to have abandoned the radical import of his own notion that the domains of technology and politics are ultimately inseparable. For the satisfaction of that desire amounts to the subordination of the technological realm to the political realm—that is, to the processes of democratic deliberation. This would restore not only the classical hierarchy, but also the much-denounced instrumental vision, for it would turn technology once more back into an instrument, this time as an agent for realizing democratic ideals.

The reason that Winner does not succeed in overcoming the dichotomy is probably caused by a remarkable inconsistency in his philosophical conception of "politics." Whereas he argues that politics is something that can also be done by things, he seems to shrink from continuing this thought when it comes to judging. At that moment, he resorts to the Jeffersonian conception of democratic politics, which is and should be done by humans alone. Thus he opens the black box of technology, but leaves untouched the black box of (democratic) politics. Had he reflected on the notion of politics, and been more thoroughly and more empirically informed about it (e.g., followed his own approach with respect to technology), Winner could have discovered that the practice of direct democracy, like technology, is not a good in itself, but should be a source of continuous consideration. In assuming that direct democracy is an unproblematic norm, Winner implies that political power exercised in this way is ipso facto beneficient, and ignores the question of how power is actually exercised in those practices.

Reflection on the democratic norm, therefore, would have to be a first consequence of taking seriously the solid alliance between technology and politics. A second consequence might be recognition of the relative independence of technologies from human intentions. Technologies, like human beings, continually resist the meanings that are thrust on them and the demands made of them; at the same time, their existence and future development depends on those who build and operate them. One might say in this respect—and in accord with the Frankenstein story—that technologies resemble nothing so much as children. Children, too, on the one hand require parental care and guidance, but on the other hand are always prone to surprising their parents and deviating from the regimen that they have been instructed to follow. Technologies, like children, once introduced into the world must be responsibly reared, without depriving them of the opportunity of surprising us, if they are to become morally and politically effective fellow citizens.

NOTES

1. See, for instance, the essays "Do Artifacts Have Politics?" and "Technē and Politeia" from *The Whale and the Reactor*.

2. This is extensively described in Maarten Hajer, *The Politics of Environmental Discourse*.

3. Stephen Cotgrove and Andrew Duff, "Environmentalism, Middle-class Radicalism and Politics," 338, cited in Hajer, 91.

4. Winner refers here to Carl Mitcham's "Three Ways of Being-With-Technology."

5. Cited in *Autonomous Technology*, 22. The original is in Francis Bacon, "Novum Organum," 537.

6. This forgetfulness is an almost literal description of the attitude associated with Borgmann's "device paradigm." In Borgmann's paradigm, technology must be instantly and effortlessly available, and the conditions of its functioning must be invisible.

7. Ellul speaks of an "ethics of non-power," in which "human beings agree not to do everything they are able to do," in "The Search for Ethics in a Technicist Society."

8. See also Rein De Wilde, *The Ironies of the Eighth Day*.

BIBLIOGRAPHY

Arendt, Hannah. 1958. *The Human Condition*. Chicago: University of Chicago Press.

Bacon, Francis. 1955. "Novum Organum." In *Selected Writings*, edited by Hugh G. Dick. New York: Modern Library.

Cotgrove, Stephen, and Andrew Duff. 1980. "Environmentalism, Middle-class Radicalism and Politics." *Sociological Review* 28(2): 333–51.

De Wilde, R. 1997. *The Ironies of the Eighth Day*, TMV Conference on Technology and Democracy, Oslo, 17–19 January.

Elam, Mark. 1994. "Anti Anticonstructivism, or Laying the Fears of a Langdon Winner to Rest." *Science, Technology, and Human Values* 19: 101–106.

Ellul, Jacques 1964. *The Technological Society*. New York: Vintage.

———. 1989. "The Search for Ethics in a Technicist Society." *Research in Philosophy and Technology* 9: 23–36.

Hajer, Maarten. 1995. *The Politics of Environmental Discourse*. Oxford: Oxford University Press.

Mitcham, Carl. 1990. "Three Ways of Being-With-Technology." In *From Artifact to Habitat: Studies in the Critical Engagement of Technology*, edited by Gayle Ormiston. Bethlehem, Penn.: Lehigh University Press.

Winner, Langdon. 1977. *Autonomous Technology: Technics-Out-of-Control as a Theme in Political Thought*. Cambridge, Mass.: MIT Press.

———. 1986. *The Whale and the Reactor: A Search for Limits in an Age of High Technology*. Chicago: University of Chicago Press.

———. 1991. "Artifacts/Ideas and Political Culture." *Whole Earth Review*, no. 73: 18–24.

———, ed. 1992a. *Democracy in a Technological Society*. Deventer, Netherlands: Kluwer.

———. 1992b. "Citizen Virtues in a Technological Order." *Inquiry* 35: 341–61.

———. 1993. "Upon Opening the Black Box and Finding It Empty: Social Constructivism and the Philosophy of Technology." *Science, Technology, and Human Values* 18: 362–78.

———. 1994. "Reply to Mark Elam." *Science, Technology, and Human Values* 19: 107–109.

Woolgar, Steve. 1991. "The Turn to Technology in Social Studies of Science." *Science, Technology, and Human Values* 16: 20–50.

Contributors

Hans Achterhuis is a professor of philosophy at the University of Twente, who writes about such themes as development aid, welfare work, and scarcity. In the area of technology studies his current interest is in the relation between time and technology.

Philip Brey is a professor in the Department of Philosophy at the University of Twente. He received his Ph.D. in philosophy at the University of California, San Diego. His research focuses on philosophy of information and communication technology, including philosophy of artificial intelligence and computer ethics.

Robert P. Crease is professor of philosophy at SUNY, Stony Brook, and historian at Brookhaven National Laboratory. His books include *Making Physics: A Biography of Brookhaven National Laboratory 1946–1972*, and *The Play of Nature: Experimentation as Performance*. He is also a columnist for *Physics World.*

René Munnik is professor of philosophy at Tilburg University and Radboud-professor at the University of Twente, both in The Netherlands. He has written a number of articles on process philosophy and on the relations of metaphysics and technology.

Martijntje Smits is a research associate in the Department of Philosophy at Twente University, where she coordinated a multidisciplinary research program on polymer waste technology. She studied chemical technology and philosophy and has worked for the national research institute TNO in the field of policy on environmental technology. Her current focus is on the philosophy of technology and technology ethics, and she is completing a dissertation on the symbolic character of the waste crisis.

Pieter Tijmes is a professor in the Department of Philosophy at the University of Twente. He teaches social and political philosophy. His philosophical research focuses on technology and scarcity.

Peter-Paul Verbeek is a professor in the Department of Philosophy at Twente University in The Netherlands. He teaches courses in the philosophy of technology and the philosophy of communication. His research concerns the mediating role of artifacts in the technological culture, especially in relation to industrial design and information and communication technology.

Index

9 780253 214492